GODS IN THE SKY

*Astronomy, Religion and Culture
from the Ancients to the Renaissance*

Dr Allan Chapman

First published in 2002 by Channel 4 Books,
an imprint of Pan Macmillan Ltd, 20 New Wharf Road,
London N1 9RR, Basingstoke and Oxford.

Associated companies throughout the world.

www.panmacmillan.com

ISBN 0 7522 6164 9

Text and line illustrations © Allan Chapman, 2001

Designed and typeset by seagulls
Printed by Mackays of Chatham plc

Jacket photographs: St. Jerome in his Study (oil on linen paper on panel) by
Jan van Eyck (*c.*1390-1441) © The Detroit Institute of Arts, USA/Bridgeman
Art Library; Hipparchus at Alexandria © Mary Evans Picture Library.

Photo credits: p1 (top) Mary Evans Picture Library, (below) Robin
Scagell/Science Photo Library; p2 Heritage Image Partnership; p3 (top)
Jean-Loup Charmet/Science Photo Library, (below) Jerry Mason/Science
Photo Library; p4 (top) Mary Evans Picture Library, (below) The Bridgeman
Art Library; p5 Steve Day/Salisbury Cathedral; p6 (top) Hereford Cathedral,
(below) Stellarvision; p7 (top) Jean-Loup Charmet/Science Photo Library,
(below) Science Photo Library; p8 Science Photo Library.

This book accompanies the television series *Gods in the Sky*, made by
Stellarvision for Channel 4. Produced and directed by Martin Durkin.

Contents

Acknowledgements

This book would never have come to be written had not Martin Durkin of Stellarvision approached me in June 2000 to discuss making the *Gods in the Sky* television programme for Channel 4. It was a subject that had long been close to my heart, and I thank him for his inspiration and for providing the wider resources which brought both the programme and the book into being. I also thank his Stellarvision colleagues Angela Neenan, Steven Green and Nicola Gillett, along with the film crew, especially Stuart Hall and Jeremy Hewson, with whom it was such fun to work. Nor must I forget all of my students over many years, on whom I have tried out and developed many of the ideas contained in this book. As far as the actual production of the book is concerned, I must especially thank Emma Tait and Gillian Christie of Macmillan, and Robert A. Marriott for his invaluable editorial assistance and indexing. My wider thanks go to the staff of the Bodleian Library, Oxford. I also thank the Revd Ursula Shone of the Society of Ordained Scientists. My greatest debt, however, is to my wife Rachel, who in addition to being a fine classical linguist, has assisted in innumerable ways, from reading and commenting upon my handwritten drafts to typing, editing and proof-correction of the finished text.

Introduction

From the most ancient times, human beings have been fascinated by the movements of the heavens. The life-giving powers of the Sun, the cyclical nature of the Moon, and the seemingly perfect and unchanging regularity of the 'stars in their courses' formed a sharp contrast with the unpredictable occurrence of droughts and floods, plagues and plenty in the world in which they lived. It was probably from these early observed regularities that the science of astronomy had its origins in Egypt and Mesopotamia in the third millennium BC.

It is my purpose in this book to attempt to trace this development, through four thousand years, down to the European Renaissance. Yet the history of astronomy is much more than a history of physical discoveries, for that emerging scientific history was always firmly embedded within parallel developments taking place in religious and broader cultural beliefs; and to imagine the growth of any science solely with reference to itself, or to other sciences, is to present a warped view of the history of thought. It is my argument that astronomy not only evolved in the past within an intimate relationship with religion and other cultural activities, but that it still does so. Just as the ancient Egyptians and Babylonians studied the celestial movements of their 'gods in the sky' and related them to wider questions of purpose and continuity in human life, so the twentieth-century realisation that the Universe began with a Big Bang begs the question of what existed before that Bang, why the ensuing Universe developed the wonderful structures that it did, and what our place is within it. And while no-one is claiming that modern astronomy, with its sophisticated techniques of investigation, must be pursued in a context of mythology and folk belief as was that of the Babylonians, we must not forget that, quite apart from the questions of beginnings, endings and purposes that inevitably emerge from it, the twentieth century has probably generated a greater volume of astronomically-related mythology than did the Egyptians in 3,000 years. Beginning with H.G. Wells' *War of the Worlds* (1898), and extending to present-day television programmes dealing with alien abductions, through a bewildering virtual universe of Greens, Greys, Daleks, crop-circles, flying saucers, pyramid alignments, *Star Trek* characters and the outpourings of Hollywood, astronomy has supplied the raw material for some of the contemporary world's most potent myths and images.

It is my suggestion that human beings are instinctively religious creatures; and if one lives in a culture in which coherent public religious beliefs are increasingly marginalised and even derided, then people make up their own mythologies from whatever material is at hand, be it a conviction of the real existence of Mr Spock, or a blind faith in a Universe that somehow or other *must* be teeming with superior intelligences. And ironically, just like all the hundreds of generations that have preceded us, it is still to the sky that modern people look for answers to their profoundest questions.

In addition to being instinctively religious, mankind is also instinctively historical, for human beings always wish to know the origins of things, from the genealogies of their individual families to the sources of those great movements which have moulded the cultures in which we now live. At the present time, when cheaper books, television, and computers have made more information readily available than ever before, the historical origins of civilisation and the great ideas that have come to frame our lives have taken on a particular fascination, as our perceived collective pasts colour our imagination and confer our identities. And just as our distant ancestors sat around their firesides and passed on tales of comets, aurorae, marriages, battles and miracles, so we, with our modern and increasingly interactive media, similarly love to explore our various histories.

Because science plays such a major part in our modern world, its history, its ideas, and the inventions arising from it have naturally become a subject of interest to many modern people. On the other hand, many of those individuals who take upon themselves the business of communicating both historical and modern science to the public frequently do so from a standpoint of what might be called triumphalist secularism. Science is not only often presented as synonymous with reason and reasonableness, but the scarcely concealed agendas of some interpreters imply that science is now the *only* mode of reasonable discourse. The assumption soon becomes apparent, therefore, that religion is *unreasonable* by definition and to be equated with superstition and ignorance, with the resulting generation of the myth that modern science only became possible after brave secular spirits dared to stand up against the official mumbo-jumbo of priests, who supposedly had some kind of vested interest in keeping 'the truth' hidden.

I believe that this image of triumphalist secularism is not only a distortion of well-documented events in the historical development of science, but also gives an ideological 'spin' to contemporary science which the evidence itself does not warrant. It is the theme of this book that without the concept of a coherently-designed Universe, made from nothing by an intelligent Deity, and contemplated and studied by human beings who possessed a particular intelligence that enabled them to unravel its structures, science as a rational understanding of causes and effects in the world would never have come into being.

It is also my argument that what was pivotal to this historical process was the rise of monotheism, or a belief in a one, singular and all-powerful God. For though remarkably sophisticated systems for observing, tabulating, and explaining the natural world grew up in polytheistic cultures, such as China, Babylon, and India, it was in those cultures that developed the concept of a single divine mind that science as an analytical, mathematical and predictive interpretation of

nature took an especially firm hold. This happened among the classical Greeks, with their philosophical concepts of the *logos* and the *nous*,
added to which was the profound influence of the Jewish One God,
which not only defined and directed the Jews' own subsequent intellectual and spiritual culture, but also that of ensuing Christians and
Muslims, who themselves adopted the belief in the One God. For by
seeing nature in all of its forms as related to one 'design principle',
rather than being subject to the caprices of various autonomous 'gods
in the sky', one opens the door to a logical, mathematical and
exploratory understanding of nature, as opposed to an essentially ritu-
alistic or placatory one.

Gods in the Sky attempts to trace this historical development.
Beginning with the cosmologies and mythologies of the ancient Near-
Eastern empires, it examines the fusion of astronomy and religion in
their cultures. Then it asks why the Greeks had such a formative effect
on subsequent Western and Near-Eastern thought. It is argued that
their philosophical belief in the above-mentioned *logos* and *nous* was
fundamental, and that when this was combined with the idea of the
Jewish One God in the works of Jewish philosophers such as Philo,
and early Christians like Origen and St Augustine, a major new direction in the progress of knowledge became possible. These ideas,
moreover, were rapidly absorbed into Islam in the eighth century AD,
to produce the world's first great flowering of post-classical science
interpreted within the context of the One God.

As these classical and Arabic ideas took firm root in early medieval
Christian Europe (for both Christians and Muslims needed a detailed
knowledge of astronomy to calculate their respective liturgical calendars), one finds the emergence of a remarkably sophisticated intellectual culture in the cathedrals and abbeys and in the newly-established
universities. And while it is true that the great majority of medieval
intellectuals were clerics, it is fallacious to accept the hoary old tale
that the medieval Church exerted a stifling effect on the life of the

mind. Quite to the contrary, indeed; for as we shall see, no medieval university student could graduate without possessing a systematic knowledge of astronomy, while certain fourteenth-century clerical cosmologists were asking questions about space and time that sound astonishingly relativistic.

The book ends with an examination of the circumstances that led to the rediscovery of what was believed to be *pure* Greek astronomy during the Italian Renaissance, and the way in which the 'new' astronomy – along with the English Bible, Shakespeare's plays and William Byrd's music – came to form a key creative component within the English Renaissance of the early seventeenth century.

However, never losing sight of the underlying theme that men and women are persistent history-making beings, we begin by looking at those very myths and assumptions that science has come to spin around its own origins, and which have produced their own particular distortions of the historical record.

ONE

Science and its modern myths

The modern world would never have come into being without science. Whether one talks about a 'pure' science with no obvious practical applications, such as astrophysics, or 'goal-directed' studies in immunology aimed at curing cancer, the power of science in the modern world is everywhere clear, both as an intellectual system and as a generator of useful technologies.

And yet, while the power of science has escalated only over the last three or four centuries, the aspiration to understand, explain, and make use of the natural world lying behind it extends back to the dawn of recorded intellectual history. Indeed, the lawyer, philosopher and visionary Sir Francis Bacon was very succinct when, in his manifesto for the 'Great Instauration' or great revival of learning in the early seventeenth century, he spoke of science being for 'the Glory of God and the Relief of Man's Estate'. It was for the glory of God insofar as scientific research opened up our minds to new wonders and filled us with a sense of awe for the God who had created them. And the relief of man's estate lay in those fruits of intellectual inquiry that could make our lives more bearable, with particular reference to advances in Bacon's favourite subjects of medicine, agriculture, and navigation. We would now define Bacon's categories as 'pure' and 'applied', or 'science' and 'technology'.

Bacon was, of course, drawing on a human aspiration that was already more than four millennia old by the early seventeenth century. For when the ancient Egyptians, Babylonians, and Canaanites (who lived in Palestine before the Jews) framed their creation stories and explained how Order had gained a precarious ascendancy over Chaos at the beginning of time, they were expressing an awareness of that complex, interrelated beauty of nature which the human mind could somehow comprehend. And by noting its cycles of light and dark, heat, drought, and inundation, they could establish those calendars and seasonal agricultural practices which are the earliest examples of science-based technology.

In this archaic world, of course, all natural forces were envisaged in animistic terms, as the products of the will or conflict of gods, and one of the watersheds in intellectual history came in the sixth century BC, when some Greeks began to think of natural objects behaving as they did because of some great underlying design, rather than because mischievous deities were being capricious with them. Indeed, whether one were trying to explain an eclipse of the Sun, the collapse of a tower, or an outbreak of epidemic disease, the Greeks were beginning to seek structural unities in nature – the presence of the *logos* or *nous*, or a logical divine agency that lay behind nature, rather than a simple chaos of spirits.

Yet to the Greeks, this logical orderly force that underlay everything, and which could be explained by disciplined human reason, was in itself profoundly divine. It would therefore be a misunderstanding of Greek cultural and spiritual priorities to interpret their realisation of nature as self-consciously secular in its intent. Religious and scientific perceptions were intimately bound together, and would remain so throughout the intervening centuries until the late nineteenth century. (In Chapter 12 we shall examine how religious thinking and scientific thinking adapted to each other after 1660.)

First, however, we shall review the ways in which modern science has constructed its own mythology, to see itself as an intellectual system or activity which could come into being only once the forces of 'superstition' had been defeated. Indeed, in doing so, science has in turn trod the same well-beaten anthropological path in describing its own particular version of the ascent of 'truth' and 'order' out of 'chaos' (or ideological error) as the Egyptians, Mesopotamians, Jews, Christians, and all those other faiths that have attempted to explain the forces that drive humanity and create its perceived place in nature. And in the process, modern science has spun about itself a myth that is no less apocalyptic and visionary than those of the religious ways of thinking from which certain apostles of science have so energetically tried to distance it.

One favourite and fondly perpetuated scientific myth – evangelised in the writings of the Victorian evolutionist Thomas Henry Huxley, and driven home in Andrew White's unequivocally titled *A History of the Warfare between Science and Theology* (1895) – is that of scientific and religious thinking as being fundamentally antagonistic. It was Huxley, White, John Tyndall and other late Victorians who, rightly impressed by the enormous strides that science had made in their lifetimes, saw it as the all-conquering force of a new order of men and women. And one has to admit that traditional Christianity – especially English low-church Protestantism, as reflected in parts of the Church of England and most of the Dissenting Churches of the early and mid-Victorian age – was temporarily stunned by science. Discoveries in geology and evolutionary biology were demanding new interpretations of the *Genesis* creation story, while increasingly powerful telescopes, discovering a wealth of nebulae and star systems, only seemed – as Thomas Hardy suggested in *Two on a Tower* (1882) – to reduce mankind to insignificance.

Yet this perceptual divide which arose between certain late Victorian scientists and the religious community was itself based as much on received cultural traditions as everything else. Staunchly Protestant England and America were deeply Biblical in their Christianity. Adam and Eve, obedience to authority, nagging concerns about divine judgment, and the imminent reality of hell fire were bred in the bone through a particularly strict and narrow reading of Scripture. Indeed, even Charles Darwin – whose own eventual agnosticism was always a thing of some regret to him – seems to have been sincerely worried for the fate of his free-thinking father's own soul, following Dr Robert Darwin's death in 1848. Because Victorian religion was so Biblical in its basis, therefore, it was especially vulnerable to shocks which challenged specific physical details in Scripture.

On the other hand, even the most affirmedly atheistical of Victorian scientists had usually been educated, and often confirmed, as Christians in childhood and youth, so that ecclesiastical images and modes of organisation often lurked in the very depths of their imaginations. As adults making scientific discoveries, therefore, they may have felt repelled from the tenets of a creed that did not seem to describe physical reality in a way in which they had come to understand it, but they not infrequently went on to turn 'science' itself into a substitute religion. And they did this by imbuing science with all the perceived hard and unshakeable truths of a dogma which could not be challenged and which could always offer convincing answers to all questions – from the predictable certainty with which the chemical elements of the Periodic Table combined with each other, to the supposedly iron laws by which the human brain reduced all emotion to neurological activity.

The assumptions and postulates of their science had taken on the status of a Holy Writ in their own right, as their search for rock-solid truth had caused them to rebound from the pulpit to the laboratory. As the prophets of a new faith, therefore, those Victorian patriarchs of the laboratory (sometimes immortalised in portraits sporting Mosaic facial hair and in postures of episcopal dignity) felt the need to widen as much as possible the chasm between their own truths and those of traditional religion. In short, they had a vested interest in taking an adversarial stance against traditional religion as a way of marking out their own intellectual territory.

What is a pity nowadays, more than a century later, is that many of their successors should be defending the same fragments of Victorian turf with the same weapons of ridicule and misunderstanding, seemingly unaware of the fact that while physicists no longer believe in billiard-ball atoms, modern religious people no longer accept as hard historical fact the ancient Jewish traditions of Adam and Eve. For why should a scientist reserve the right to change his or her mind about a once partial understanding of a scientific truth, while prohibiting a religious person the right to bring new light and interpretation to an old spiritual truth? Could it be that the creation myths of modern science are just as intolerant of alternative ways of thinking as many modern apostles of 'scientism' like to imagine those of theologians to be?

Indeed, over the last hundred years – and especially over the last fifty or sixty years – religious understanding has developed no less rapidly than has science itself, and not infrequently this has happened by both science and religion working cooperatively, so that one is nonplussed to find some modern-day scientist-critics of religion still stuck in an 1860s Darwinian controversy time warp. For while it is true that science can never prove the factual accuracy of specific points of religious faith, some sciences can cast valuable light upon the context of religious events. Archaeology, for instance, has revealed information about many Near-Eastern cultures that has provided illuminating background understanding of Egyptian and Babylonian myths and Biblical narratives. Likewise, psychology and anthropology have enabled us to find remarkable parallels between creation stories and the psychology of religious belief from numerous cultures around the globe. And that analysis of ancient texts which began with Victorian Egyptologists, Assyriologists and Hebraists, and which now makes use of the most sophisticated computing technology, has made it possible to trace earlier derivations and identify original additions to ancient documents. In 1872, George Smith's discovery and translation of the Assyrian Gilgamesh epic – found in the library of King Assurbanipal (c.650 BC), but probably dating back to the second millennium BC, prefiguring as it does the later Jewish narrative of Noah's Flood and Ark – was one of the first great surprises of scholarly research in this field, indicating that the *Genesis* narrative had not been dictated by God to Moses intact, but was perhaps influenced by an earlier source. Indeed, these discoveries were all related to the broader movement called 'higher criticism', which was concerned with subjecting Christian texts to an intensity of philological and historical scrutiny similar to that applied to the surviving documents of secular history and literature. And no-one can deny that the ensuing discoveries could be profoundly destabilising for those Victorian Protestants whose faith lay in the assumed premise that every word and letter present in the Bible was absolutely true on a simple and obvious level of understanding.

Yet what all of these discoveries and new insights have done is to force religious people – in particular, Christians and Jews – to reassess the grounds on which their faith stands. It has made them realise that instead of religious truth dropping out of the sky in one perfect whole, as had been assumed to be the case with the Biblical Pentateuch (the first five books of the Old Testament), spiritual perceptions have passed through many channels, insights, and inspirations borrowed and incorporated, with profound truths and genuine revelations mixed in with legendary material – mythological material, indeed, that seems less convincing to us nowadays because it is garbed in the clothes of an archaic culture, the peoples of which saw the world in terms very different from how we see it today. Yet these stories are still capable of conveying deep spiritual truths when interpreted from a standpoint that is meaningful to us in the twenty-first century.

In this way, for instance, the simple yet dramatic *Genesis* narrative of the creation of the world can be understood within the context of evolution. For what both the ancient Jewish story and its twentieth-century reinterpretation share is a sense of awe at God's creative power and His ability to develop and sustain a world of intense beauty and complexity, and to have a conscious relationship with beings made in His own image.

In short, religious thinking has become flexible and exploratory, and to think of it as trapped within a straitjacket of ancient prescription which renders it innately antagonistic to science is truly a myth of the age in which we live. In fact, Darwinism was not only a wonderful discovery for scientists, but also became one for theologians. Both Aubrey Moore and Frederick Temple (who later became Archbishop of Canterbury) pointed out that evolution had *not* shown God to be a remote and aloof deity who had simply set the Creation running at some stage in the distant past, and had then perhaps left it alone (as the deists claimed), but displayed him rather as an active participator who constantly added new wonders to His original design.

A second myth of modern science warmly perpetuated in certain quarters by 'evangelical' atheists and agnostics is that all religious belief is essentially fundamentalist in character, in contrast with the presumed evidence-driven flexibility and openness of science. Yet, as we saw above, contemporary Christian and Jewish thought about modern science is refreshingly aware of new intellectual currents, and realises that God speaks to us in a language that we can understand within the context of the age in which we live. In the same way that one would not explain natural phenomena to a twenty-year-old adult in the same terms as one would to a child of four, so we accept that a creation story intended to be meaningful to a people whose geographical knowledge extended little beyond the Sinai desert could not be adequate for people today.

Of course, those scientists who have a vested ideological interest in perpetuating the divide between science and religion, and who lack the curiosity and intellectual flexibility to explore what modern-day religious people really think, find it very convenient for polemical purposes to paint their perceived foes as simple fundamentalists.

Ironically, however, it is often religion's scientific critics who most often behave like fundamentalists today. For their fundamentalism resides in a simple belief in the wonderful and unlimited power of science to achieve almost anything, when once stripped free from 'spiritual superstitions'. Scientists of this type, for instance, will frequently go to the most Byzantine lengths of speculation – extending well beyond the experimental evidence – to avoid the slightest hint of divine or any other deliberate purpose in nature. One encounters cosmologists who will use their computer models to twist theoretical protostates of helium and hydrogen through every permutation of implausible fantasy rather than countenance the merest possibility that perhaps God preceded the Big Bang.

Some sciences – such as cosmology, and neurological studies of consciousness – are in fact not just straightforward descriptions of nature, but inevitably pose questions about the beginnings, endings, and meanings of things. In short, they involve ideas touching upon philosophy and theology. Yet when scientists who study these challenging and thought-provoking subjects refuse as a point of creedal allegiance to consider, even hypothetically, any form of non-physical explanation in the understanding of nature, then they cease to display that intellectual openness which all scientists generally accept as the hallmark of their profession. They then slide into what might be called a rigid materialist fundamentalism; and ironically, when this happens, their own materialist fundamentalism becomes as blinkered, exclusive, and puritanical as that of the religious bigots whom they despise, as one of science's myths about open-mindedness is found to double back upon itself.

Of course, what science's myth of undogmatic openness shares with religious systems going back to the dawn of recorded history is a concern with purity. In the same way that the ritual priests of Ra, Anu, Marduk or Yahweh needed to follow specific prescriptions of thought and conduct if they were to separate themselves from the rest of contaminated humanity, in order to serve their god and fulfil their priestly functions, so too those modern scientists who insist upon a purely materialist explanation of all phenomena feel that they must not contaminate themselves with 'superstition' or 'religion', or even philosophy, if they are to be fully effective in their vocation. Needless to say, this narrow material puritanism is by no means the only approach to science, for there are, after all, devout Jewish, Christian, Muslim and other scientists involved in front-rank research at the present time who do not feel the need to be material reductionists. But nonetheless, the materialist model has now come to enjoy enormous prominence, especially in the media, and many people are led into the assumption that real scientists and real intellectuals do not need the divine.

When defending their materialist stance, however, those scientists who think this way often create another myth of modern science which itself has powerful religious connotations: namely, that even if science cannot explain a particular favoured scientific model today (for example, a purely brain-centred explanation for consciousness), it will eventually be able to do so in 10, 50 or 100 years' time. Yet by allowing themselves to argue in this way, scientists often take such imprudent leaps in the dark within the intellectual assumptions of their discipline, that to an outsider it all appears astonishingly like blind faith in the future!

The materialist may exclaim that the history of the last 500 years has shown science making relentless progress in the development of physical knowledge. And it most certainly has; but often by moving in very different directions, and ultimately reaching very different conclusions, from those predicated by the original scientists who first pointed the way. Victorian physicists, for instance, believed that because Newton's Laws of Gravitation had been accorded their comprehensive expression, the behaviour of all matter was, by and large, explained, with one time–space constant permeating the vastness of space. But by 1930, Einsteinian relativity and quantum physics, with their bent and warped space and strange relationships subsisting between matter and energy, had changed these earlier and apparently conclusive physical perceptions beyond recognition.

By 1900, the leading cosmologists in Europe and America believed that the available weight of observational evidence indicated that the thousands of misty nebulae (later called galaxies) visible, through powerful telescopes, in the depths of space, all belonged to one single, stable, all-encompassing star system that was an extension of the Milky Way. After 1923, however, Edwin Hubble's discoveries were to turn that view completely about, as his new data demanded a radical reinterpretation of former certainties. Hubble's observations indicated that the nebulae were separate, distant, 'island universes' in their own right, and appeared to be expanding in all directions into the very depths of space, instead of being fixed and relatively local.

Nor are the life sciences free from fundamental switches of apparent direction. The eighteenth-century Swedish botanist, Carl von Linnaeus, had, by 1778, shown from extensive and careful field researches that all living species were fixed and incapable of crossbreeding. Then in 1859, Charles Darwin produced, from the fruits of new original researches, an opposing interpretation, indicating how, across a vast time-scale, species evolved under natural selection. And later in the 1860s Darwin's cousin, Francis Galton, put forth the seemingly reasonable suggestion that a carefully devised plan of selective breeding of humans would eradicate weak strains from the population and increase the overall percentage of the clever and the fit. It was a well-intended suggestion – based on the best biological evidence of the day – for a society worried by the apparent escalation of paupers and vagrants in its midst, and seeking a way to a more just and prosperous future. Yet Galton could never have envisaged how 'eugenic science' would be developed by the Nazis during the 1930s and 1940s. And who today would really defend the so-called impartial scientific basis of Sigmund Freud's model of the human mind, with its oedipal deter-minism and its attempt to explain most human actions in terms of infantile sexual traumas?

Of course, no-one is suggesting for one moment that science, as a progressive, forward-moving way of explaining the natural world, cannot change its course in the wake of new evidence; for any disci-pline that stands on an ingenious reformulation of questions to nature, the framing of new experiments and the invention of new techniques of analysis and interpretation, must inevitably change its direction.

On the other hand, once this point is acknowledged as a general principle, it soon becomes impossible to predict in detail the direction which science will take and exactly what it will achieve during the decades hence. Who in 1900, for instance, would have imagined that pulmonary tuberculosis would have been largely banished from the Western world by 2000, whereas the common cold would have remained as elusive as ever? Or that we could put men on the Moon, yet still be incapable of developing a safe and reliable railway system?

To say what science will achieve in the future, or what problems it will solve in ten or a hundred years, therefore, presumes an expec-tation on the part of the individual that is beyond the evidence avail-able at any one time. In short, it predicates a myth about the nature of progress, and presupposes a leap of faith in what science *will* do some time in the future.

There is also another way in which scientific knowledge – like all other forms of rational knowledge – hangs upon a leap of faith. For as the Scottish free-thinking philosopher David Hume argued in his *Inquiry into Human Understanding* (1748), sensory knowledge can never reveal the true causes of things, only their effects. When a hammer hits a bell we always hear a clang, but we can never know the true cause of the clang. We can only say that every time in the past we have seen a hammer hit a bell we have always heard a clang, and that the blow and the sound have a 'necessary connexion'. Though, as Hume realised, the mental habit of associating a blow with a sound works as a way of explaining daily experience, it is not a *proof*, but only a '*belief* [that] is felt by the mind'.

This might appear to be like splitting hairs, but in essence it means that when we talk about ordered-sense knowledge – such as logically connecting the results of an experiment with the processes that we observe within the experiment while it is taking place – we once again need to make a leap of faith. On a very profound level, therefore, the scientist must have a faith in the fact that acids will always turn litmus paper red, in the same way as the gardener has faith in his seeds to produce plants of a specific type. We must assume, in other words, that the observed 'necessary connexion' that has always linked A with B in the past will always do so in the future.

As Hume was aware, this lack of ability to prove the causes of things not only struck a blow at the method of science, but also demonstrated that it was no more susceptible to solid proof than were the deductions of religious argument. Of course, neither scientists nor theologians, nor any other logical thinkers, can make any progress if they become too worried about their inability to prove the connectedness of knowledge. Life, after all, has to go on, and ideas have to be developed. Science is thereby reminded that it does not possess an unchallengeable remit to use its method as the sole arbiter of all truth, and that scientific as well as religious knowledge is ultimately unprovable.

Another of science's favourite myths about itself is that it is impartial and above opinion: science is 'true' because it is not 'prejudiced'! Yet while the vast and mute structures of nature have, one supposes, a separate existence and innate logic from us human observers, science itself is *not* nature. Science, rather, is a system of investigation which aspires to explore nature's inner logic, but which is itself invented and managed by fallible men and women. And like all things managed by mankind, it will become the victim of choices, tastes, and prejudices. Since at least the seventeenth century, experimental science has advanced not through the rigid application of a 'method' – no matter how much working scientists may declare their admiration for Sir Francis Bacon or René Descartes – but by the playing of hunches. In the same way as a cross-examining lawyer like Bacon played his hunches when breaking down a villain in the dock during a Jacobean treason trial, so an ingenious, lateral-thinking scientist decides to follow a specific experimental procedure when investigating a particularly elusive piece of a natural phenomenon. Indeed, the possession and strategic use of the creative 'killer instinct' – be it in law, science, or any other logical activity – is what separates the geniuses from the standard practitioners. It is the use of this 'killer instinct' which makes possible fundamental discoveries in science, as one scientist sees possibilities missed by everyone else.

For centuries, for instance, scientists had known that a slab of glass could be made to produce a spectrum of colours when sunlight was passed through it. And from Aristotle onwards, it was assumed that this took place because the pure celestial light of the Sun was 'decayed' by contact with a terrestrial substance. It took the inquisitive, lateral-thinking mind of the 24-year-old Isaac Newton to change that view between 1666 and 1669.

Using a pair of glass prisms purchased (it is thought) at Stourbridge Fair, Cambridge, Newton first used one prism to produce the familiar colours. Then, using his second prism to pick up the coloured light exiting the first prism, he found that the spectrum could be recombined into white light. And if he went on to pass rays of each individual colour of light, from red to blue, through a pinhole in a card and then through the second prism, then the colour exiting that second prism would not break down further, but always remain the same. From these simple experiments he drew the correct conclusion that white light is a combination of what he generally regarded as six primary colours; and in doing so he changed the entire course of optical physics. Yet that creativity, or instinct, about how to obtain and interpret evidence, was not obvious, or even straightforwardly logical. Rather, it was the result of that package of hunches, musings, lateral mental flashes, and willingness to examine received opinions afresh, peculiar to the mind of Isaac Newton, which made modern optics possible.

Crucial experiments, indeed, are rarely ever self-evident either in their devising or in their interpretation, even when made with familiar objects. For no less than poetry or painting, creative science hinges on individual peculiarity and insight – and this, by definition, governs the perception and choices of direction made by the experimenter in pursuit of his research, in very much the same way that Vincent Van Gogh perceived the world very differently from John Constable, although both men were creating great art.

So while science has evolved techniques and procedures for studying nature – such as the preference for the simplest possible explanation for a particular phenomenon – the application of these procedures when faced with particular circumstances in the real world is by no means automatic. The scientist must make choices at every turn, and have 'faith' that a specific mode of procedure is correct, and will be intellectually valid and genuinely informative. And these choices, moreover, extend well beyond the ways in which one may set up a laboratory experiment to work within a particular set of parameters. The choices also depend on whether an individual scientist is, or is not, sympathetic to 'religious' explanations. Is he or she likely to be inspired by the seeming design present in nature, and see no objections – at least in principle – to considering God as the creator? Or does that scientist have, for whatever reason, what might be called an ideological aversion to religion, and instead feel a commitment to a materialist, reductionist, position?

Either way, we must never forget that the choice is not inevitably disclosed by the results of an experiment, for while nature's processes contain an undoubted inner logic, nature herself is remarkably silent about her origins, purposes and moral intentions. Nature simply exists. It is only in the sensoria of our minds that we think about purposes or lack of purposes, and the forces motivating an individual to one or the other opinion probably has more to do with how we frame our private value systems than it has with any rigorous proofs about the status of the external world.

Therefore, as the ultimate truth of neither religion nor science can be absolutely proven and must be taken on trust as a form of *belief* – as David Hume realised nearly 250 years ago – it would appear that a scientific materialist mode of explanation depends just as much upon faith as does that of divine creation. The key difference is that religious people in all traditions fully accept the 'faith' status of their beliefs, arguing rather that scientific experiment is not the *only* and *exclusive* pathway to knowledge. Reductionist scientists, on the other hand, often work on the assumption that their own position is grounded upon a form of self-evidently observed fact, and in denying that all knowledge involves a component of faith, envelop themselves in one of the most pervasive mythologies of the twentieth century: namely, that religion is about the unprovable, whereas science deals with basic hard realities.

Yet is faith, by definition, 'blind'? It can be, most obviously when associated with some irrational superstition such as, for example, the inveterate gambler's stroking of a 'lucky' charm prior to purchasing a lottery ticket, but not when faith is invested into something with an established track record of success. Faith is not blind, for instance, when it produces consistent results. For in spite of Hume's caveats about connections and not knowing ultimate causes, faith in experimental scientific knowledge does produce reliable results on a practical level in the real world. Who, for example, can deny the reality of the science that underlies the atomic bomb, modern medicine, or space travel?

While religious fruits are less physically tangible than is a penicillin injection, they can nonetheless be equally transforming, for faith in God has dramatically changed both individual lives and the fate of nations. Religious faith can, and often does, provide a sense of coherence, understanding, and acceptance to those for whom a faith in medicine has failed, and who yet have learned to face death with a tranquillity of spirit. And would those great humanitarian movements, such as the abolition of slavery, or the foundation of hospitals and refuges for society's most economically unviable members, ever have taken place without religious faith, demanding as they did real and actual sacrifice on the part of the powerful in response to a spiritual command to assist the poor and outcast? In these respects, therefore, faith in religion shows itself as capable as science of bearing real and tangible fruits that change the human condition. Although the specific expectations from investing faith in religion are different, such faith nonetheless has an equally profound transforming effect upon the world in which we live, and is therefore not blind, but based upon a foundation of reason.

One might well speculate, therefore, as to the factors which incline an individual either to a mythology which acknowledges the primacy of the divine, or a mythology of self-sufficient matter. One suspects that personal temperament and circumstances play a much bigger part than any rigorous proof: good or bad religious experiences, whether one is drawn by disposition to clean and tidy, perhaps minimalist, solutions, or whether one rejoices in the complex and multilayered, or whether one has an attraction to or a repugnance from explanations in which matter and spirit intermingle.

But the scientific determinist may respond to the above by saying: 'Is not science proved to be true on the very strength of its achievements in explaining nature and creating the modern world?' The results of modern science are often used to lend weight to the myth of science's ultimate truth status, but this myth can be countered by two arguments.

Firstly, science's wonderful power to explain phenomena and to change the circumstances of life need not, as we have seen above, be inevitably married up to a model of exclusive materialism. To say that God initiated the Big Bang, set going that creative tension between orderly and chaotic forces which lies at the heart of evolution, or gave us those 1,400-gramme packages of dopamine and circuitry that we call brains to generate consciousness and ask the profoundest of questions, in no way diminishes the wonder of the actual science. The insistence upon a materialist solution to these questions is in fact motivated just as much by culture and attitude as is our taste in art or music. Indeed, I find it odd that some neurologists, whose intellectual passion is the study of our brains, will perform somersaults to avoid even considering the attribution of *purpose* to the most sophisticated problem-solving structure that we know to exist in the length and breadth of the Universe. After all, if our brains are programmed to search for purpose and order, why should we feel so ideologically compelled to deny their very existence? The acceptance or rejection of purpose, therefore, is not an inevitable consequence of scientific research, but is the personal philosophical choice of the individual scientist.

Secondly, why is it that, after science has done so much to transform the modern world, there is a significant minority of people within the West who seem to be disillusioned by it? Could this have something to do not with science's ability to cure diseases and produce labour-saving technologies, but with the attitudes displayed by some of its leading practitioners? Could it have something to do with the gospel of determinism and big power preached in certain quarters? Could it have something to do with science's seeming willingness to rubbish the spiritual, downplay the needs of the individual, and even – as is the case with some contemporary prophets of science – treat those who disagree with them as simpletons?

No-one, of course, is suggesting that scientific truth should be judged by its popularity value in a plebiscite. Honest intellectual endeavour, pursued with techniques of the highest sophistication, cannot be judged by the same standards as a television game show. Yet if prominent scientists who are active propagandists for their discipline project to the public an aura of condescension and disdain in which their own personal contempt for the spiritual becomes associated with the basic attributes of the discipline as a whole, then many people may feel repelled by science. And this is all the more sad when we bear in mind that these attitudes are in no way an inevitable corollary of science itself.

When one looks at science, and how it has developed during and since the seventeenth-century scientific revolution, one finds that many aspects of both its values and its methods stem from what might be called religious concepts. One of these beliefs is that there is a fundamental order in the Universe and that this order can be explained by human intelligence. Indeed, without this assumption, neither science nor any other ordered mental activity could exist. (Some of the order versus chaos mythologies will be discussed in Chapter 2.) Another is the Platonic idea that the highest level of human knowledge is the understanding, through geometry, of the perfect and eternal *forms* present in the 'mind' (for want of a better word for a non-personalised force) of the *logos* or *nous*. Without it, indeed, science would have lacked the necessary intellectual schema to develop geometrical and logical proofs in antiquity, and the physical laws of motion in the scientific revolution and thereafter.

Similarly, when science sees itself as a force for human good, as 'relieving man's estate' through medicine or technology, it touches upon ideas of moral imperative which first arose in the religions of the ancient Jews and Egyptians, who saw themselves as protected by wise parental deities. This instinctive tendency to personalise or attribute moral properties to aspects of the natural world or cosmos not only survives into our own day and age, but even lurked, for at least three centuries, in the minds of people who consciously disavowed religion.

The tendency to personify nature and deify 'her', for instance, is found in many of the freethinking philosophers of the Enlightenment, while more extreme figures such as Julien Offray de la Mettrie wished to free human thought in general, and science in particular, from what he perceived as the dead weight of traditional Judaeo-Christian religion, arguing that human beings were nothing more than 'self-acting machines'. 'Nature' – now with a capital N – came to epitomise truth, reason, and justice, especially in salon society in mid-eighteenth-century France. Jean Jacques Rousseau's fictional hero in *Émile* (1762) grew up happy and strong, influenced only by the loving forces of 'Nature', and innocent of formal education or organised religion. 'Nature' also came to be associated with optimism and the prospect of a Golden Age, and when the fanatical ideologues of the French Revolution abolished the Christian Church in France in the early 1790s, they replaced it with festivals of Reason and Nature. But the rational regeneration of mankind failed to halt the butchery of the Reign of Terror in 1794.

Yet even when traditional religion was formally opposed on intellectual or political grounds, it was quickly replaced by new systems that were religious in all but name. For in their concerns with reason, order, law, Nature, rational morality, and a new, perfectible, and distinctly apocalyptic view for the future, the freethinkers of the

Enlightenment were still bound on a subconscious level to the Christian mental habits of medieval and Renaissance Europe.

It could even be argued that our present-day concern with human rights, beginning (as mentioned above) in the late-eighteenth-century campaign to abolish the slave trade, is profoundly indebted to religion, even when the bases of those rights are expressed entirely in secular terms. For in a truly dog-eat-dog world, lacking any kind of spiritual mandate, and in which the only practical value of human intelligence is to outwit and overcome our human competitors, human rights have no place. Indeed, this starkly savage 'State of Nature', in which life was truly 'nasty, brutish, and short', was explored to its limits by one of the founding fathers of the Enlightenment, Thomas Hobbes, in his *Leviathan* (1651).

To escape from this original barbarous condition and enjoy the peace and protection bestowed by civilisation, said Hobbes, our pooled resources and rights of personal defence must be laid at the feet of an absolute and arbitrary ruler who can do whatsoever he likes except kill us. Yet in spite of its pugnacious secularism, *Leviathan* bristles with Biblical quotations and ideas, as Hobbes – the son of a Wiltshire parson, and an Oxford graduate – attempts to find a workable basis for a materialist political and social ethic. This ethic, moreover, was to be built upon a 'science' of mental geometry, where matter in motion explained not only the processes of the natural world and the workings of our bodies, but also the perceptions and pre-programmed instinct of our own minds – such as the instinct to triumph over our opponents by whatever means came to hand, and to avoid death for as long as possible. Hobbes became notorious in his own time and thereafter, by arguing that in a purely material world, no-one had any natural human rights, because there were no absolutes beyond physical survival and the laws of matter.

And now, more than three centuries later, our modern concerns regarding human rights are still framed in the light of the Beatitudes in the Sermon on the Mount, although we often try to ground them in a secular ethic. It would appear, therefore, that one of the foremost moral questions of our time – the basis and nature of human rights – has its roots firmly in religious thinking. For without our myth of human specialness, grounded long ago in our perceived relationship with a Supreme Being, it is difficult to determine from where arose a rational awareness of our 'rights'.

It seems, therefore, that whether we are in Heliopolis at the height of the radical Akhenaten revolution in Egypt around 1550 BC, in Paris in 1792, or in London in 2001, human beings are compulsive myth-generating creatures, and builders of those transcendental systems which we might call religions. And perhaps the greatest irony of all is that we still build these systems even when hotly denying the real existence of religion itself. For as we have seen above, science has both historically, and also today, drawn heavily on such religious concepts as order, predictability and intellectual coherence in discussing the structure of the natural and moral worlds, and in our attempts to understand them. Yet what I would also argue is that modern science has constructed such a framework of reference for itself as to give it many of the characteristics of a coherent religious mythology in its own right. The parallels between modern science and traditional transcendent religion, therefore, are interesting to explore, and can, I believe, be set forth as follows.

One of the most obvious ways in which science takes on the aura of a religion is in the sense of powerful and even arcane knowledge possessed by the scientist. The possession of this knowledge sets the scientist apart from the rest of common humanity, for he or she now has powers and insights unique to an order of priests. Science becomes a triumphalist body of knowledge, defeating all that stands before it, and enforcing its own rules by which to judge what is true and what is false, in much the same way as the post-Exodus Children of Israel invaded the land of Canaan and tried to eradicate its ancient gods. And as is often the case with an evangelical religious faith, science is frequently depicted by its most fervent advocates as working only when taken as a *whole* package, for one is strongly urged to accept the materialist philosophy along with the explanations of how chemical compounds are formed and how the wonders of digital technology work.

In addition to the triumphalist ethic, scientists, like priests, must go through a long exclusive training, part of which hinges on learning the arguments and interpretations on which the individual's faith is founded and why, indeed, it is the true one. Science also has its own shrines where truth is revealed, such as laboratories, and a hierarchy of cognoscenti, extending down from Nobel Prize winners to skilled laboratory technicians. One might also suggest that the ubiquitous white coat is the instantly recognised vestment of the order.

One of science's most potent mythologies is about its own origins. Just as the ascendant and all-powerful Marduk overcame Apsu, Tiamat, and other Babylonian chaos gods after periods of trial and struggle, so science depicts itself as finally breaking free from the chaotic forces of superstition. And like the Christian Church out of which modern science was born, science has its calendar of saints. For in addition to its self-generated and self-perpetuated myth of the warfare between science and religion, mentioned above, nineteenth-century science in particular developed its own hagiography.

Firstly, there was Saint Christopher Columbus, who supposedly contradicted the doctrines of a blind and bigoted Church by suggesting that the Earth was round instead of flat. In spite of the purely mythological status of this belief (for every educated person, whether priest or layman, knew Ptolemy's arguments for a spherical Earth by 1492), it has proved to be amazingly resilient, and in particular, many Americans over forty will recall how, when in junior school, they were told of the brave Italian navigator who had discovered their country because he dared to challenge the dogma that the Earth was flat. (Columbus and the influence of his discoveries will be discussed in Chapter 9.)

Perhaps a less popularly familiar saint of science is Andreas Vesalius, the Belgian anatomist who worked at the University of Padua in the sixteenth century. He had dared to go against the Church by dissecting human cadavers, and had discovered thereby several structural details in human bodies which 'dogma' said should not have been found there. Yet what Vesalius contradicted was not the Church, but the conservative interpretations of his fellow doctors who insisted on the gospel status of the works of classical Greek anatomists such as Galen, but who had in fact misunderstood how the human body works.

Then there was Saint Nicholas Copernicus, who was so afraid of the all-powerful Church, with its thumb-screws and red-hot irons, that he did not dare to publish his great discovery that the Earth went around the Sun until he was on his death-bed. This is, of course, a fundamental distortion and simplification of the truth if ever there was one, for Copernicus was a Canon of Frombork Cathedral, Poland, a deeply pious Christian, and a friend of bishops and cardinals (as we shall see in Chapter 10).

Then there was Saint Giodano Bruno, a later follower of Copernicus, who really was burnt at the stake in Rome in 1600. But he was *not* burned for his purely scientific beliefs, but for his religious pantheism, which saw God as a remote and aloof Creator, and hence cast doubts on the divinity of Christ, and indeed discounted the need for the Church altogether. And when one remembers that all of this was being argued by a man who was a Neapolitan Dominican Friar who had also worked as a Protestant spy against fellow Catholics, one can understand that Bruno was perhaps as far out in his heresy as one could reasonably go at that time, and his Copernicanism paled alongside the magnitude of his political and theological heresies.

And did not Saint Galileo Galilei at his enforced recantation of his Copernican views before the Roman Inquisition in 1633 courageously whisper under his breath, 'Nevertheless, it *does* move'? But here again, the popular 'Galileo versus the Church' hagiography entirely misses the point that Galileo was, like Copernicus, a pious Roman Catholic, and that his heresy was a technical administrative one that was brought to a head by Galileo's partiality for ridicule, and the crossing of a frankly over-mighty Pope who had once been Galileo's friend. Politics and personal pride figured more prominently than science in Galileo's condemnation.

Yet what the scientific hagiographers invariably fail to do is remind their readers that in condemning Galileo, the Church quickly realised that it had shot itself in the foot, and no more Copernicans were punished. Indeed, subsequent Jesuit astronomers, such as Giovanni Baptista Riccioli and Roger Boscovich, openly discussed Copernicus's ideas as a *theory* (as opposed to a *fact*) in the context of Ptolemaic orthodoxy, and got away with it (see Chapter 10).

In addition to these 'saints' in the scientific calendar, there was also Charles Darwin, who undoubtedly caused some ripples within the Victorian Church of England. However, as we saw above, contemporary theologians and scholars came to terms with evolution surprisingly quickly, seeing within it a wider and more awe-inspiring sense of the divine. Yet, one may argue, by saying that men and women were descended from apes, was not Darwin undermining Adam and Eve? Not really, for while Darwin's work could not be reconciled with a simple literal interpretation of *Genesis*, there was no reason why human beings could not be understood as creatures whose bodies had evolved by natural processes, but whose self-aware-ness and propensity to ask profound questions about the beginnings and endings of things – our immortal souls – had been implanted in them by a Creator.

Indeed, nothing brought home the existence of this popularly perceived 'saints' calendar' of scientific martyrs more forcibly than an invitation which I received some time ago to participate in a local radio programme dealing with aspects of the history of astronomy.

Going over the outline of the script with the producer immediately prior to a live transmission, I found that all the hoary old myths about the liberation of astronomy from Church suppression were present, and the producer was mortified to hear my 'revisionist' remarks, which upset the whole planning of the programme. So Copernicus was *not* punished by the Church for saying that the Earth went around the Sun? So the Church did *not* have an express policy of suppressing science? So Giordano Bruno was burnt for heresies that were *theological* and not *scientific*? And Galileo was *not* a courageous unbeliever who was crushed by the Church because he dared to challenge its blind, ignorant, dogmas? As one might imagine, the planned message of the programme was knocked sideways. When I asked the researcher where she had got these dinosaurian tales of science, I was told 'off the Internet'. So beware!

Perhaps because all of the world's religions are aware that they contain mythological components, people who deal with religion – such as theologians – have not only a sensitivity and a respect for the transcendent, but a realisation, that as human experience changes, so we sometimes need to modify those myths through which we try to define the transcendent. Religion, in consequence, is instinctively good at modifying the details of its myths without losing the big picture. Science, on the other hand, because of its aspirations to impersonal objectivity and an often dogmatic assertion that matter is all that exists, can sometimes be blind to its own internal mental habits. And when it does this, science forgets that, as a human intellectual creation, it too contains its own inner mythologies. It is particularly sad, moreover, when these mythologies seem to project an image of science which is bleakly deterministic, for this only has the effect of driving people away. And it is tragic when people feel driven from science to seek explanations involving *unreason*, because they feel that such explanations – astrology, magic, Earth-worship, New Age occultism, along with the fashion for emotional wallowing

– give them more of a sense of personal power or control over their lives than does 'big science'. (These modern movements are further discussed in Chapter 12.)

It is the argument of this book, however, that science not only grew out of religion, but that, as human knowledge became broader and more sophisticated over time, the two developed a creative inter-action with each other, each casting light upon different aspects of human experience. For science and religion not only spring from the same intellectual roots in the human psyche, but also stand on very similar ground today. For both science and religion stand for *reason*, as opposed to *unreason*; for the belief that *truth* really exists and can be arrived at by disciplined modes of thought, as opposed to merely being a social construct (as post-modernism would like us to believe), and that *order* not only has a higher truth status than *chaos*, but can lead to a better state of existence, both in this world and in the next. Yet we must never forget that while science offers us a world of wonders and useful things, it is by no means a neutral value-free system, but contains within it, indeed, many mythologies all of its own.

And at the present time, when the relationship between religious and scientific ways of thinking is arousing more interest than it has done for more than a century, and Heisenburg's uncertainty principle and chaos theory are opening up new possibilities to physicists and philosophers, it is salutary to remember that one of mankind's most ancient concerns was that of trying to understand the connection between order and chaos.

TWO

Order, chaos and cultures

One of the earliest intellectual problems with which the human race had to contend was in determining the reasons for change. What are the sources of order and chaos? And why is human life so precarious, with famine, disease and accidents everywhere to be found; whereas the heavens, by contrast, seem to display an eternal regularity? It was the peoples of the Near East, some time after 3000 BC, whose explanations of these problems – expressed in the form of creation, destruction and regeneration myths – came to have an enduring influence on the Western mind, for their myths encapsulated a philosophy of nature that attempted to find rational explanations for why things behave as they do; and in some cases, how human actions might alter them.

The influential Near-Eastern cultures were those of the ancient Egyptians, Mesopotamians, Canaanites and Jews. For one thing, they had several features in common, such as rivers that flooded, a close proximity to deserts – which made the contrast between the ordered and the chaotic seem quite stark – and climates that could be capricious, sometimes bringing drought, and sometimes destructive floods. These cultures, moreover, spanned a zone of latitude between 25 and 33 degrees north, which was sufficiently close to the equator to make it possible to see astronomical constellations that were quite deep down below the celestial equator, while being sufficiently northerly to see the stars of the north circumpolar region. At the same time, these cultures were all close enough to each other to have regular mutual contacts both commercial and military, which no doubt facilitated some of those exchanges which resulted in some of their mythologies sharing common themes.

However, at the present day, when our knowledge of the natural world has been transformed by experimental scientific methods, one might argue that archaic creation myths have nothing to teach us. And while this is true when it comes to explaining the physical processes which we now know govern the world, the myths represent, nonetheless, the first documented attempts to form a rational understanding of things – rational in the sense that archaic creation stories embody connected sequences of ideas which, within their own system of rules, produce solutions which are both relevant and satisfying to the cultures that produce them. And while it is true that our knowledge of the details of these myths, and of the cultures that produced them, has increased enormously in the wake of nineteenth- and twentieth-century archaeological and philological discoveries, some of the stories and the characters they contain have been known, at least fragmentarily, for centuries through the Jewish Old Testament, and the compilations of Roman writers such as Plutarch and Cicero, who were fascinated by the ancient gods and their ancestries.

Some of the oldest and most powerful of these myths about creation, order, and chaos come from Egypt. Across the three millennia that preceded the Christian era, the official theologies of the priests of Heliopolis, Memphis and Hermopolis produced variations in these myths, and sometimes changed the status of the individual gods within them, but the broad patterns of the stories remained remarkably consistent. Indeed, consistency and continuity are the hallmarks of Egyptian culture across the board, and it is a mistake to see these characteristics as signs of mere conservatism. To the Egyptian mind, stability was a defining precondition for a balanced and just world in which chaos was kept at bay, and in which gods, acting through the divine intermediary, Pharaoh, and people lived in harmony.

This necessary condition of stability, however, was to have a long and tenacious history in the rational explanation of human and natural phenomena. Greek medicine was to interpret health as a balanced state of hot, cold, moist, and dry humours, while political philosophers down the centuries were to define a good state as one in which all competing demands were held in equipoise. And while modern science sees the natural world as a place of constant change, from the evolution of living things to the expansion of the Universe, we still see those things as taking place within the context of general laws. The mathematical bases of the DNA code, gravity, quantum physics and relativity all enshrine those constants without which organised knowledge would be impossible. The cosmologies and cosmogonies (or origins of the gods) of the Egyptians and other ancient peoples, therefore, represent the very genesis of a cultural trait which is still a cornerstone of modern intellectual endeavour.

Where, of course, the ideas of the ancient cultures differed most noticeably from modern thinking was in how the forces of order and chaos were visualised. For with their very different views of the natural world, the people of 3000 or 2000 BC saw natural agents – from the motions of astronomical bodies to the visitations of fevers – as animistic in origin. A god or demon must be at work. Yet in an ordered state of nature even these animistic forces were not to be viewed as irrational in their operation, for just as the Egyptian god Osiris could be seen as making things go right, so Apophis could be guaranteed to make things fall apart. In its own way, therefore, the

Egyptian cosmology was just as rational and predictable as ours, for the contending forces could be counted upon to act in certain ways.

It was in everybody's interest to placate and keep on the right side of Ma'at, the goddess of order. The priestly rituals in the temples were to ensure the maintenance of Ma'at, and when this was success-ful, the Nile would flood copiously and at the right time, the heavens would be constant, and peace and justice would prevail in the land. Indeed, when a new Pharaoh came to power, he was traditionally presented with a statuette of the goddess Ma'at (recognisable by her ostrich-feather headdress), and in the Hypostyle Hall of the Temple of Karnak at Luxor, one sees a Pharaoh being presented with a stat-uette of Ma'at as part of his royal commission to maintain order and *cosmos*. But if Isfet gained the upper hand, then the reverse would be the case, as an anarchy in the divine and in the natural worlds threat-ened order. Yet these two personified forces of Ma'at and Isfet were as eternal as everything else in ancient Egypt, and their relentless interaction was used to account for the often unstable world about us. Even so, Ma'at always seemed to win, for the world, the heavens, and the gods continued to exist. Later ages would think of these forces as daring versus fortune, Christ versus Satan, order versus entropy, and, more colloquially, as success versus 'sod's law'.

None of the archaic Near-Eastern cultures, however, had a concept of creation *ex nihilo* – or creation from nothing. Such a concept was to be a product of post-Davidic Jewish thought of around 1000 BC, as enshrined in the Biblical *Genesis* narrative. Instead, they all saw water as the starting point. To the ancient Egyptians this was Nun, the primordial boundless watery chaos that preceded the world. Creation happened when, so the Heliopolis cosmogony said, a co-existing yet sleeping and inert demiurge somehow made a cry. A primordial hill or mound now appeared above the waters of Nun as the process of creation began. It became Egypt, Heliopolis, and spread out to create foreign, and by definition inferior, lands and peoples, for to the Egyptians they themselves were not only the primary race of mankind but indisputably the superior one.

The demiurge called Atum in the Heliopolis cosmogony created the first Egyptian gods, such as Shu, god of air, and Tefnut, goddess of moisture. Light, astronomical bodies, and *cosmos*, or unity, were now summoned into being. Over the millennia, and in different religious centres within Egypt, different gods were seen as the primary agents in creation. It was sometimes said that Ra – who came to be seen as the Sun god – was the source of all things, while the theologians of Memphis viewed the parent god as Ptah, the god of the Earth. Either way, the creation took place at an unimaginably remote time in the past, and represented not just Ma'at versus Isfet on an adversarial level, but also something more philosophical: the triumph of coherence above incoherence, and being over non-being. Its subliminal themes of potential and actual were to find resonances in the thoughts of Plato's pupil, Aristotle, whose whole explanation of rational thought and natural change hinged on the doctrine of the potential becoming actual, interpreted by 350 BC in terms of logical syllogisms. And just as the creation took place at an indefinably remote time in the past, so that creation would last for an unspecified series of ages into the future.

One of the central astronomical myths of ancient Egypt was that of Nut, the sky goddess, whom the Greeks were to equate with Rhea, their own sky goddess. Nut was the twin sister of Geb, who was also seen in Heliopolis (but not in Memphis) as the Earth god. In what would become an Egyptian royal custom, Nut and Geb married incestuously, but because they did so without the permission of the all-important Sun god, Ra, they were afflicted with barrenness, as Ra made it impossible for Nut to give birth during any of the months in the archaic 360-day Egyptian year. But Thoth, the god of record-keeping, calculation, calendars, and orderly practices in general (whom the Greeks later identified with Hermes, the god of speed and intelligence) had pity on Nut's plight. Amongst Thoth's duties was a responsibility for the Moon – a fact borne out in Egyptian art by his being sometimes depicted as wearing a crescent head-dress. According to the narrative later recorded by the Greek Plutarch, Thoth then engaged the Moon in a series of games of draughts, and as he was on a lucky streak, he succeeded in winning $\frac{1}{72}$ part of the Moon's annual light, or a total of five new days, which were inserted in the old calendar before the New Year. This resulted in a year that now contained 365 days, and gave Nut five days in which to give birth to Osiris, Horus, Seth, Isis and Nephthys. (One wonders whether this myth was developed as a way of incorporating the realisation into Egyptian astronomical knowledge that the year contains 365 days rather than 360 days. For a pre-mathematical culture which would have found it difficult to detect any sensible changes in the height of the noonday Sun when it 'stands still' for several days in succession at both the midsummer and the midwinter solstices, this would have not been easy to discover.) Nut is often depicted in Egyptian religious art in the form of a slim young woman who arches across the Earth, so that her body forms a canopy above it. She is also frequently shown with stars painted upon her, to represent the vault of the night sky. One of the most beautiful depictions of Nut – with both the Sun and also the stars on her body – is on the painted ceiling of the tomb of Pharaoh Rameses VI in the Valley of the Kings near Luxor.

Considering the geographical location of Egypt, however, there would have been several natural factors that contributed to the Egyptians' belief in their own cosmological specialness. For one thing, their country must have appeared as a long, thin fertile groove cut through the desert, and irrigated by the annual inundations of the Nile. On both sides of this luxuriant, long oasis was barrenness, occupied by dangerous creatures, and beyond it there were only foreigners. Egypt, moreover, ran very largely from south to north, almost down the meridian, so that when one looked south, astronomical bodies always rose to the left from the desert, culminated to the meridian above the Nile, and set to the right, beyond the western desert. And was not the life-giving Nile itself reflected in the very heavens themselves, in the form of the Milky Way? And over the course of the seasons the Sun-god Ra made his long and short daily progress across the sky, passing through Nut's starry body and being endlessly reborn.

Ra's daily journey across the sky was performed in a boat, from which he spread light and warmth over the world, and ordered Ma'at. At night, however, Ra and the lesser or companion gods who journeyed with him in the Sun boat would sink, old and tired from their day's celestial voyage, into the underworld. But their subterranean nightly journey back to the eastern sky was not a time of rest, but of struggle against Apophis, the blind, senseless serpent-like god who embodied Isfet, or primal chaos, and whose only apparent skill lay in his ability to scream and cause pandemonium. Each night, Apophis tried to block the course of Ra's Sun boat, but with the help of his fellow voyagers, including Osiris, the god of regeneration, various Protectors, Seth (now, apparently, reformed in later mythology from being a chaotic to a preserving deity), Thoth, and others, the boat always successfully passed through its Twelve Gates, or stations of the underworld, and after the correct rituals had been performed, reached the eastern horizon intact. There, the voyagers were reborn, young and fresh, to flood the world with dawn light, the rosy colour of the dawn sky being equated with the childbirth blood lost by Nut as she gave new life to the Sun. Nut, we must also remember, was the mother of Osiris, who was the first of her five children, born supposedly in Thebes, over the five extra days in the year won from Thoth in the games of draughts. Osiris had fallen foul of Seth when, one presumes, Seth was still a brutal, loud spreader of chaos, and before he came to take his position as a 'reformed' god on Ra's Sun boat. At this earlier stage in his existence, indeed, Seth had murdered Osiris and cut his body into pieces. His wife and sister Isis, however, assisted by mother Nut, put him back together again, after which he became god of the underworld.

Only in a timeless cosmos, of course, could order fight with chaos and always win – a fact clearly witnessed by the repeated rebirth of Ra and Osiris each morning. Similarly, only in such a cosmos could gods be regenerated like Osiris, and even in some cases the same function, such as charge of the Earth, be interchanged between two separate characters, as was the case of Geb in the Heliopolis cosmogony, and Ptah in that of Memphis. But while all of these deities could act within human time – as witnessed by their activity at particular times of the day or night – they really existed in a time frame of their own.

The relationship of the Egyptian gods with human beings, however, was generally benign. Humans were part of that richness of creation, and they were connected to the divine through the Pharaoh, who himself had a divine descent from Ra. And when a dead Egyptian passed to the underworld, he or she could expect, if they had not added to the amount of Isfet in the world, to enjoy an eternal life similar to their earthly life, stretching into a limitless future. And even though peasants would still have to work in this future state, they would at least do so free from pain and old age. Indeed, it is in their view of an afterlife that the Egyptians perhaps differed most from their Near-Eastern and even Graeco-Roman neighbours, most of whom envisaged the world to come as a dark and rather forbidding place.

Central to Egyptian cosmology was the temple. Unlike the church, synagogue and mosque in later centuries, however, the temple was not a place of public worship for ordinary people. Temples, such as those at Heliopolis, Karnak, and Thebes, were, rather, living visualisations of the primordial hill or mound that emerged from the waters of Nun at the beginning of worldly time, and represented models of creation itself. They were cosmogonic foci, often inhabited by their patron deities, and needing attendance in the form of prayers and sacrifices performed by an exclusive priestly order, if they were going to maintain Ma'at, and keep the Sun rising and the Nile flooding.

The cosmology of the Egyptians was one of the most humane in the ancient world. Perhaps this was engendered by the relative physical security of the Nile valley from outside attack, combined with the very regular flooding of the river, which gave agriculture an enviable predictability. Very clearly, the religious liturgies worked, and hence served a supremely rational and practical function in that society. Myths, after all, generally reflect the way a country feels about itself, and on the whole Egyptian mythology is about stability and well-being.

This was not the case with the peoples of the Tigris and Euphrates valleys, however, where those cultures known as the Sumerians, Babylonians and Assyrians developed very different cosmogonies and cosmologies. Several factors, indeed, could have helped mould their myths, which were often of a dark and threatening nature. For one thing, the climate and general physical environment were more unforgiving than those of Egypt. The flooding of the rivers was much less predictable in terms of water volume than was the Nile, and was less rich in nutrients, while the water did not always come when it could do most good for agriculture. Because of the topography the land was not, generally speaking, covered with a thick, rich coat of fertile mud, as in Egypt, but required the digging and maintaining of miles of drainage channels to divert the erratic water supply on to the parched fields. Mesopotamia's generally flat terrain, moreover, endured an essentially continental climate, in which hot, dry winds and uncertain rain made life hard for farmers; and because of the relative lack of big natural obstacles – such as the harsh desert that surrounded Egypt – it was always prey to invasion and political unrest. Hard labour in an unforgiving, unpredictable environment produced in the Mesopotamians a perception of those 'gods in the sky', who controlled all things, which was very different from that of the Egyptians. And while details vary between ideas developed by the early walled city states of Sumeria in the south of Mesopotamia and the later empires of the Babylonians and Assyrians to the north, the basic details have an essential consistency that spans more than two millennia.

As with the Egyptians, all things were seen as having come from water, for in the beginning there was only a limitless sea of fresh water, called *apsu*. Then somehow this ocean begat the Earth and the sky, and a demiurgic force drove them apart, to produce land, sky and water.

Anu, or Anum, was the chief god in the pantheon of the ancient Akkadians of Uruk, and he seems to have been followed by Enlil, representing, respectively, the gods of sky and Earth. Anu was more remote, and controlled that realm in which the astronomical bodies moved, and while Enlil was a god of earthly providence, he could also be treacherous. Enki (known in later texts as Ea) emerged as a sort of civil service agent for the greater gods, and Enki, among other things, had superintendence of the waterways and drainage canals. Ishtar, the sister of the Sun god Shamash, had the supervision of storms, while Ninurta, the son of Enlil, was responsible for social harmony.

Around 2000 BC in southern Mesopotamia, the religious understanding of these peoples saw order as maintained by Enlil through the keeping of the Book of Destinies. This universal record, which foretold everything that was going to happen, was the instrument by which Enlil retained his power over the lesser deities. Always lurking and awaiting his opportunity to overthrow the divine world, however, was Anzu, a predatory vulture-like bird-god, who on one occasion succeeded in stealing Enlil's all-powerful Book of Destinies and regalia when Enlil was resting. Enlil then became ineffective, a council of the gods was convened by Anu, and the champion Ninurta defeated Anzu and retrieved the Destinies, after which Enlil was able to regain his authority and rule as before. In another chaos narrative, the sea monster Labbu, who lurked in the *apsu* ocean that ran under the Earth, was defeated by a storm god, and order was restored.

Truth and order, therefore, come to be seen as residing in stability and hoped-for changelessness, although they are threatened by ill-meaning spiritual predators. But peace is always restored through success in combat, thereby reinforcing the importance of success in war as the obvious way in which to maintain stability. Indeed, one can understand why military aggression was bred into the bones of these Near-Eastern cultures, for war not only defended the land, but also reinforced spiritual harmony.

The Babylonian epic poem written on seven clay tablets, known as *Enuma Elish* (from its opening words 'When on High'), however, presents a later and more elaborate cosmogony. In the beginning in *Enuma Elish*, the fresh waters of the primordial ocean, now more clearly personified in the male god Apsu, mixed with those of the female salt ocean of Tiamat. The other gods and goddesses were formed from this water, but at this point a new principle was introduced into the cosmogony which constituted a major source of disturbance: movement, with its ensuing noise. With this noisy end to tranquillity, Apsu threatened to kill his and Tiamat's divine children, for he could no longer sleep, but Tiamat colluded with their son Enki, or Ea, to put the old god to sleep, kill him, and take possession of his powerful regalia. War now broke out amongst the gods. Emerging from this chaos, however, was Marduk, the favoured grandson of Apsu, who even as an infant god had been given the storm-winds as a toy to play with, and he became the champion of order. Destroying his grandmother Tiamat, he cut her body in two, and from the pieces, and the waters that gushed forth from them, Marduk created the world, the heavens and the stars.

Marduk emerges, therefore, as the supreme creator god in the Babylonian pantheon, representing order, energy and potential, and his temple in Babylon comes to be revered as the place where creation began, and where he lives. But all of this took place in the period before men and women were created, for in the hierarchical Mesopotamian pantheon, once the world had emerged from its primordial *apsu*, the gods had to meet their own needs, keep the irrigation ditches open by their own efforts, and, presumably, look after themselves. And while it is true that murders could take place, generally speaking the gods were immortal beings who seemed to live in a dimension that was uniquely their own, and not in the same time-frame in which mere humans would come to live.

And then the gods decided to invent mortal mankind – effectively as a labour-saving device. Now these transient beings would do the hard work, as a subject people living at the bottom of the cosmological heap. Temple tributes were needed to keep the gods well fed and comfortable, and in return for all the hard work, people hoped that the gods would at least not punish them unnecessarily. Indeed, it was a singularly harsh view of the divine in which no quarter was expected or given. In one account, in fact, Enlil even threatens to destroy the human slave race because people have become so numerous and noisy, just as the lesser gods had once been, and he cannot sleep; but the other gods protest that he spare them – not from compassion, of course, but because the lesser gods realised that they themselves would have to go back to work if the people were destroyed.

Although the gods were the only immortals in the respect that they supposedly retained their identities and status for ever, human beings never actually died. Instead, after earthly death, an individual person passed to a dark, bleak, and awful place ruled over by Nergal and his Queen Ereshkigal. This was indeed a radically different after-life from that of the Egyptians, although it has some parallels to the realm of Pluto in Greek mythology.

The Mesopotamian creation narratives, and the relationships perceived to exist between the gods themselves and between divine and human beings, reflect in many ways the insecurities of the environment from which they came. One can also understand how divination came to play a crucial part in the civic life of the Mesopotamians, and with it the study of the movements of the stars and planets, made, as they were thought to be, by Marduk from Tiamat's body. Could one indeed use the observed regular motions of these divine fragments to establish a foundation for order? It is interesting to note that, in the Babylonian state in particular, astrology was invented for political purposes and the maintenance of civic stability, though from it grew the personal horoscope that could be of use in guiding individual lives. In Chapter 3 we shall see how Babylonian astronomy worked, and how their 'gods in the sky' were found to embody order and lay the foundations for the use of number and proportion as a way of studying the heavens. For by the time that Babylon finally fell in 538 BC, it had produced an astronomical culture of such sophistication that even the early Greeks stood in awe of it.

As we have seen, conflict myths were an essential part of cultural identity and cosmological meaning for both the Egyptians and the Mesopotamians. They were also to play a major role in the cosmogonies of those who dwelt in the Biblical land of Canaan. Resonances of these myths echo through the Jewish Old Testament, for the Canaanites and the inhabitants of Ugarit had originally occupied what later became the land of Israel long before the Jewish monarchy established by Kings Saul, David, and Solomon began to change the intellectual and spiritual history of the world around 1000 BC.

The Canaanites worshipped an all-powerful creator god named El. He was so supremely powerful, however, that his actual contact with the world seems to have been distant, as he lived in a tent at the centre of the Universe. But El had a capable lieutenant in Baal, who appears to have had the job of supervising the Universe on more of a daily basis. Baal, who is the false god of the ancient Jews, vilified by the prophet Elijah and mentioned in the Books of *Kings*, *Numbers*, *Judges*, and elsewhere in the Old Testament, had a sister called Anat.

Like all the Near-Eastern deities, Baal had to be especially vigilant in his control of those two perennial sources of chaos: flood and drought. It was Yam who brought flood, and Mot who brought drought. And in the best traditions of Near-Eastern divine war, Baal's sister Anat was also a fighter, for like Isis in Egypt, Anat brought her brother back to life when he had been killed by Mot. Baal, however, was the great governor, and through war as well as through good celestial government the order of the cosmos was maintained by him.

All of the gods mentioned above, however, exist in pantheons which have important things in common. Firstly, they are all staunchly ethnocentric deities, concerned with a specific group of people whom they will protect if they feel like it (which means that they like the sacrifices and temple liturgies performed in their honour), and punish if they feel neglected or if the whim takes them. This provides a valuable cultural tool, for it means that whatever happens, one has a seemingly rational explanation for good and bad harvests, storms, plagues, and plenty, and can also make sense of the environment. And if these deities can treat their own people roughly, it is anyone's guess what they can do to foreigners!

A surprisingly large number of these mythological systems, moreover, saw the world as coming into being from a primordial and inert body of water which a demiurge principle brought to life to initiate creation. But to essentially arid regions where, instead of rain, it is the hoped-for seasonal inundation of a great river that brings new life, this is perhaps more understandable. And as water creates life, so it can take it away, as was the divine punishment sent to the world in the Babylonian flood epic of Gilgamesh, probably dating from the second millennium BC, and in *Genesis* in the time of Noah and his Ark.

And while the theogonies of Heliopolis, Memphis and Hermopolis in Egypt, and those of Nippur, Uruk, Babylon and Ugarit to the east provided genealogies for their gods, we must not forget that the jumbled relationships that sometimes existed between them were the results of trying to describe in time events and personages that preceded and no doubt would succeed human time. The chronologically bizarre relationships between Ra, Nut and Isis, and Anu, Tiamat and Marduk must not be seen as simple linear narratives. Instead, they were attempts by peoples living 4,000 years ago to make sense of order and disorder through a series of stories that were fragmentary in their archaic oral origins, and continued to be couched in animistic terms after they had been tidied up by scribes and written down, perhaps centuries later.

Indeed, resonances of these ideas were to be found in the later Greek theogonies of Homer and Hesiod, while in the sixth century BC Thales was to see water as the primary agent from which all things come. And like the earlier Near-Eastern peoples, the Greek philosophers were motivated to find order and meaning in the world and to explain why things changed.

It can be understood that as these mythologies developed over millennia, and as centres of power shifted, the power-gods of an older culture would be seen as growing old and being superseded at the front of people's minds by younger and more vigorous ones. This was especially important in the crucible of Mesopotamia, where war, and possibly climate changes, saw a gradual shift of initiative to the north, from the older, southerly Sumerian city states to the much more centralised empires of Babylon and Assyria. Yet as Marduk emerges as the all-powerful governor-god of Babylon by around 1800 BC, older Sumerian cities like Nippur still acted as cult centres for Enlil and the older gods.

What all of these cultures further shared was the belief that the cosmos and the first gods who fashioned it were preceded in time by something: Nun, or Apsu, or the formless watery chaos. However, in the first millennium BC, and probably originating in the second millennium, a major innovation was made by the Jews: namely, that God was one and singular, and that he created the heavens and the

Earth from nothing – or, as it was later expressed in the Latin, *ex nihilo*. Yahweh had no pantheon of lesser deities, and was seen as existing both before time began and as succeeding it after He eventually chose to wind up and end His creation. And as He made the world from nothing, so that the whole creation was the product of one supreme intelligence, the cosmos was capable of being in harmony with itself. This creation narrative, immortalised in the Biblical Book of *Genesis*, was destined to become one of the world's most far-reaching and intellectually influential theogonies, and for several reasons.

Firstly, if Yahweh had fashioned everything from nothing, or *ex nihilo*, then the matter out of which the Universe had been made need not be 'recalcitrant', or resistant to a perfect design. This recalcitrance, indeed, was a problem which would later trouble Plato, who saw creation in terms of an intellectually perfect potter trying in vain to impress his perfection upon the recalcitrant clay which he had *not* made *ex nihilo*, and which therefore proved resistant to his modelling.

Secondly, the *ex nihilo* creation gave Yahweh a special relation with the human race. As there were no ranks of bickering lesser gods to cause trouble then the human creation became the sole focus of Yahweh's concerns. Instead of Enlil, Nut, Marduk or Baal, there are only Adam and Eve. And what is of crucial importance in the history of ancient cosmogonies, Adam and Eve were unique in having been made in Yahweh's very image, so that men and women resembled God Himself. Adam was moulded first from the dust of the earth by Yahweh's own hands, and then Eve was fashioned from Adam's rib after God had caused a 'deep sleep' to fall on Adam. (This passage, in *Genesis* 2, later led the Victorian Scottish surgeon Sir James Simpson to point out that God Himself had been surgery's first anaesthetist!)

Adam and Eve, moreover, enjoyed a unique distinction in the cosmos, not only because their bodies were fashioned by God's own hands, but because the breath of life which made them uniquely human and different from the beasts had been breathed into their nostrils from God's own mouth. This made humanity potentially immortal, and gave it a fundamentally different status from that envisioned by other Near-Eastern cultures.

In the Garden of Eden, over which Yahweh gave Adam and Eve dominion, one finds an earthly place unique in the literatures of antiquity: a paradisiacal place, wherein there is no disharmony or conflict. This Eden, moreover, seems to be without seasons or weather, where the lion ate grass with the lamb, and where an eternal spring seems to have prevailed. Indeed, seventeenth-century chronologists, such as Archbishop Ussher and John Lightfoot, even attempted to fix the exact day and hour at which the world had been created, from *Genesis* and from other Hebrew written sources.

And as Yahweh had made everything from nothing, and separated the light from the darkness on the fourth day of His six-day Creation, to produce the Sun and the Moon, the stars, and the rest of the Universe, one had an explanation for the apparent eternal perfection of the celestial realms through which ran the astronomical bodies.

But this perfection is not the world that we know. For along with that power of reason and self-knowledge which God had breathed into them with their immortal souls, Adam and Eve were tempted to pry into God's innermost knowledge, to unlock the very secret of creation. And when Eve was beguiled by the serpent (that forever-lurking harbinger of sudden death, in all Near-Eastern countries, and a mythical creature of destruction prefigured in Apophis, Tiamat, and others in the creation narratives) to eat of the Tree of Knowledge, the supposedly innocent couple suddenly found that their eyes had been opened and they had glimpsed God's knowledge, which is an understanding of evil. Adam and Eve were now driven out of Eden by an angel brandishing a flaming sword, into a place beyond the garden. Here, their original sin of disobedience and curiosity had condemned them to a life of pain and hardship. Of their first two sons, Cain murdered his brother Abel, and for this he was cursed by God, though still allowed to live, for he and his parents were still special to Yahweh.

In the strictly monotheist Jewish creation story, chaos comes into the world as a result of mankind's own fault in disobeying God and eating the forbidden fruit which gives knowledge of good and evil. With this sin, one presumes, the paradisiacal Eden disappears, lions now eat lambs, and mayhem and destruction come into the world. Indeed, medieval Christian theologians, who accepted the *Genesis* story as a literal narrative, even argued that the Moon's phases were produced by Eve's sin, which was, after all, cosmological in its impact, travelling up as far as the Moon. The resulting phases made it the only astronomical body visible to the naked eye to undergo regular physical changes. Yet these changes were not chaotic, but proceeded with exact predictable regularity, making the Moon the natural barrier which separated the eternally perfect heavens from the sin-infected 'sublunary' regions of earthly change.

After Eden, Yahweh continues in what appears like a love–hate relationship with His creation, rather like the loving yet exasperated parent of a perennially delinquent family. Yahweh will punish transgression, allow His children to be stricken by famine and plague, and be carried off into captivity by their enemies, but in the end will always deliver them – for as the Jews who worshipped Yahweh always knew, they were His 'Chosen People'.

On the other hand, as even a cursory reading of *Genesis* reveals, there are components in the stories which do not tie up as neatly as one might expect. If, for instance, Adam and Eve had only one surviving son after Cain murdered Abel (*Genesis* says Adam and Eve had Seth and daughters subsequently), where did Cain's wife and mother of Enoch come from, after Cain had settled in the land of Nod? And why should a perfect creator, who presumably possessed a knowledge of all things to come, have set a subversive reptile in his garden? If it was to test the free will which he had bestowed on Adam and Eve, why was the failure of that test followed by such cosmologically potent consequences?

What we see in the Jewish creation story, of course, is an attempt to narrate creation from a new set of first principles, wherein a sole and all-powerful deity develops a relationship with His human creation which is unique in the ancient world. Even so, modern scholarly analyses of *Genesis* and the other early books of the Old Testament suggest that the Near-Eastern Semitic nation which became the Jews had earlier and more polytheistic creation narratives. And while those earlier Canaanite, Hittite, Babylonian and other narratives were expunged from the new Yahweh-based monotheism of the first millennium BC, their traces are still to be found in the so-called Mosaic writings. And as nineteenth-century Biblical scholars came to realise – especially after ancient Near-Eastern languages were gradually liberated from their cuneiform scripts after about 1860 – the Jewish creation and historical books were not pieces of unified authorship. Moses as an individual, it was realised, had not written *Genesis*, *Exodus*, *Leviticus*, *Numbers* and *Deuteronomy* – the five books of the Pentateuch – by acting as God's scribe, as he led the children of Israel into the promised land of Canaan following their release from Egypt. Instead, the Pentateuch and other books of the Old Testament were shown as having been compiled and edited from earlier sources, rather than being singular compositions. As the surviving clay tablets from the libraries of Hammurabi, of the eighteenth century BC, and elsewhere were translated, the names of ancient pagan gods, stories of their exploits, and the creation narratives discussed above, began to come to light.

Deities such as Baal, for instance, who made appearances as impotent heathen gods or idols in the Old Testament, suddenly received a wider context in pre-*Genesis* Near-Eastern creation stories. Characteristics attributed to Canaanite or Babylonian gods were also seen as having been attached to Yahweh. For example, El – the tent-dwelling supreme god of the Ugarit peoples – was referred to as the 'Most High', which was to become one of Yahweh's titles. In the *Psalms*, moreover, Yahweh was variously described as displaying the characteristics of a storm-god who 'came flying upon the wings of the wind' (*Psalm* 18), as the punisher of the great sea-serpent Leviathan (*Psalm* 74), and as the supreme deity who 'judgeth among the gods' (*Psalm* 82). In the beginning of the book of *Job*, furthermore, Yahweh seems to be on quite friendly conversational terms with a character which the English Bible styled as Satan, but who could well have had his ancestry in a much older Near-Eastern deity. Indeed, it was during God's strolling conversation with this Satan figure that the very decision to tempt pious Job to blasphemy was taken.

It is not known exactly when the Jews developed that monotheism which made their faith unique in the ancient world and produced a creation narrative of such intense power. One certainly finds Yahweh's uniquely powerful presence as the great driving force behind the achievements of King David around 1000 BC, and the *Psalms*, which were either written by him personally or else were posthumously dedicated to him, elevate Yahweh as a supreme creative force, while other gods are mentioned as false or evil beings. In David's time, however, polytheism and the worship of Canaanite and other deities continued in the land of Israel. One of the reasons, for instance, why the reign of David's son, King Solomon, was an ultimate failure derived from Solomon's toleration of heathen worship, especially that practised by his numerous foreign wives. In the First Book of *Kings* (chapter 11), Solomon even built 'an high place for Chemosh, the abomination of Moab, in the hill that is before Jerusalem, and for Molech, the abomination of the children of Ammon'. And one subsequent king of Israel after another came to grief because of his failure to put a stop to pagan practices amongst his people. Indeed, it may not have been before the seventh or sixth century BC, when the later sections of the Book of *Isaiah* were written, that the full monotheistic glory of Yahweh was finally established, and the creation story given its definitive cast.

Does this mean, therefore, that modern scholarship has fundamentally undermined the plausibility of belief in a divine creation story? I do not believe that it need affect it at all. For if one sees a creator God as involved in a constant process of revelation, why should He not have revealed Himself and His purposes by degrees across human history, in the same way that humanity's understanding of the Universe has deepened from the earliest mythologies through Ptolemy, Newton and Einstein?

One might take as an example the Book of *Job* itself, which contains the most beautiful astronomical imagery in the Old Testament. After reminding Job of his own limitedness (chapter 38), and asking 'Where were you when I made the Heavens and the Earth?', God goes on to remind Job of several classes of natural phenomena for which there were no physical explanations in the first millennium BC. Where had the Universe come from? How had Orion, the Pleiades, Arcturus and other astronomical bodies been set in the sky? What lay in the depths of the ocean, and what were the sources of lightning, hail and weather? And by what processes did living things gestate and come to life?

And yet, one might rightly argue, modern science provides us with genuine explanatory insight into these supposed mysteries, through astronomy, oceanography, geophysics, meteorology and biology. But what the sciences actually reveal to us are the wonderful *processes* by which nature works. Science, we must never forget, helps us to understand *effects* and *processes*, not *causes*. For while we, unlike Job, can now use a pressurised submersible vessel to enable us to pass through the ocean depths to understand its wonders, and follow the life process from conception to birth with ultrasound technology, we still cannot explain *why* things are as they are. Indeed, science can tell us nothing about *why* things happen; and in this respect, Job's willingness to bow down before a vastly greater power of understanding than his own is just as appropriate a response today.

There are several areas in which the Jewish creation story was to become central to the development of the Western mind, and establish a vital foundation upon which modern scientific knowledge could be built. The most basic of these derives from its emphasis that the whole of nature – both cosmological and biological – is the product of one single, unified, creative intelligence. I would argue that not only is modern science itself a 'monotheistic' structure in the respect that one body of coherent intellectual principles are seen as running through the whole of it, but that such a unitary view of nature would not have been possible without a prior cultural predisposition to a 'grand design'. The explanation of natural forces in animistic terms, in which particular deities fight with each other, might produce a cause for flood and famine that is 'rational' within the assumptions of a particular culture, but that 'reason' generally leads us to see human beings as the victims of capricious gods; and an axiomatic capriciousness is not a foundation upon which a *scientific* system of reasoning can be built.

In the Jewish monotheistic tradition, however, human beings are not victims. As possessors of immortal souls and creatures that stand in God's image and partake of His mind, they are more like children or perhaps students. Yes, the blight of Adam's and Eve's original sin of disobedience and excessive curiosity brought home their limitations when it came to understanding the ultimate purposes of things, yet as keepers of God's garden they had a custody of nature that made them unique in the history of human ideas. For that custody, combined with that power of understanding the *processes* (if not the beginnings and endings) of nature given to mankind by God, created the ability – as the seventeenth-century German astronomer Johannes Kepler expressed it – to 'think God's thoughts after Him', and thereby think our way through nature's divine blueprint.

While it is true that this set of ideas did not create a 'scientific revolution' in ancient Palestine, nonetheless, when this view of creation was absorbed into Christianity, and early and medieval Christian Europe became heirs to Greek mathematics and philosophy, a vital precondition for a scientific view of nature had come into being. And early Islam, which was also heir to Jewish monotheism and Greek philosophy, also developed a powerful scientific tradition which reached its zenith around AD 1000, although it never went on to develop a full experimental tradition as was the case in the West.

Another factor – which in addition to the possibility of unity of design and of understanding that Jewish monotheism bequeathed to the development of a scientific view of the world – was that of a coherent time-scale. For if the creation of the world also set nature's great processes into motion, and these processes possessed a predictability that derived from the divine mind, then time *began* with creation. In consequence, history itself, and the subsequent account of all that would happen thereafter, had to take place in time. This results in a view of nature which is radically different from that of the other religions of remote antiquity. Within this monothe-istic creation there can be no jumble of squabbling deities doing things *out* of time, or mixing up the sequence of events to produce worldly chaos. Thoth, for instance, cannot win a part of the Moon's light to create extra days within existing time; nor can the gods constantly kill and revive each other, or fight every night with evil forces which they always overcome.

The time frames of the Egyptian and Babylonian religions were essentially circular, with no beginnings or endings, only endless ritual circularity. The Jewish creation, however, was set within a strictly linear time-scale. Not only did God create the world; it was ostensi-bly intended to go somewhere and then stop. It is true that this jour-ney was not intended to be scientific, but it was certainly sequential. There was a Creation, a fall, an attempted reconciliation between God and his children, and guidance through the Jewish Law, along with prophets, kings and captivities. And then God himself came into the world in Jesus Christ, only to be rejected, and the Gospels, letters, and prophetic writings that His first-century Jewish followers wrote about Him saw these events as implicit within the prophecies of Isaiah from the eighth to the sixth centuries BC. The Judaeo-Christian writings – especially books such as *Daniel, Enoch*, and *Revelation* – also foretold an end to the world, and a separation of humanity into those who would be saved and those who would be lost.

Jewish chronologers and visionary writers – especially in the early Rabbinic period of Judaism in the second century BC, when Jewish persecutions had become intense in the Near East – came to look for signs and Messiahs whom God would send to liberate Israel. These visions of liberation and judgment often contained sequences of coded numbers or 'times', along with secret seals that had to be broken or keys that had to be found – as in the visions of *Daniel* – which set later generations of Jewish and Christian scholars hunting for clues and sometimes performing fantastical calculations in their attempts to find the end of time. This computational concern also carried over from Judaism into early Christianity, the origins of which were thoroughly Jewish, for Jesus Christ's first mission as Messiah had been to gather up the lost sheep of Israel. But as Christianity spread into the Gentile or non-Jewish world, this preoccupation with calculating the time of Christ's return to judge the Earth and go on to end the physical cosmos spread with it. This 'pursuit of the millennium' or of the beginning and ending of time initiated a cultural concern with sequential history and attempts to 'model time', which we express nowadays in the context of trying to model such things as the Big Bang. We now ask: will the Universe expand for ever, or after reaching a certain point, will it collapse in upon itself? But to think of matter as possessing a predictability which coherent scientific laws can express, one first needs a firmly established cultural belief in the directional character of, or at least the sequential reliability of, time, and the capacity to ascribe beginnings, middles, and ends to observations and thought processes.

The cultures of remote antiquity developed remarkably powerful ideas about circular and directional time in the context of their religious traditions, to form satisfying explanations for order and chaos in the world; but how, in fact, did they conceive that the Universe was actually built, and what did they think it looked like?

THREE

The three-decker Universe: the ancient arrangement of space

While the ancient creation stories and explanations for order and chaos tell us a great deal about how people believed that their cultures related to nature and to the divine, an understanding of their astronomy and cosmology opens up the question of how they believed that the Universe was constructed and arranged in physical terms. All of the Near-Eastern cultures shared a series of basic physical concepts when it came to their visualisation of the Universe, and what they all had in common was a cosmos that was believed to be arranged in a series of flat planes or a vault.

Generally, there were three or more such planes, so that one might visualise the cosmos as being similar to an old four-poster bed, with a mattress (representing the Earth), an over-mantle (the sky), and a chamber below (the underworld). And also like the over-mantle of an old bed, the sky was kept in place above the Earth by supports.

Egyptians, Mesopotamians and Jews all saw the Earth itself as a flat, usually circular, plate. At its remote edges were thought to be mountains or pillars, commonly described as being four in number, and representing, thereby, the 'four corners of the Earth' which held up the sky. The Jewish *Psalms* (especially 75) and the First Book of *Samuel* speak of them, while the Earth itself has secure foundations 'that it should never move at any time'.

Above the world was the sky. The Egyptians generally envisaged this as a sort of vault that perhaps, at its most far-flung extremities, bent downwards to meet its supports, although the Mesopotamians and Jews generally spoke of it as being flat. Across it the gods in the sky moved, only to pass each night beyond the pillars into an underworld which extended beneath the Earth, in order to reach the eastern sky to begin a new day.

Although the Babylonians tended to speak of their cosmos as comprising *six* layers – three for the sky, and three for the world and underworld – the general threefold division held for most cultures until some time after the sixth century BC. The *Enuma Elish* creation narrative provides three: Heaven, Earth and Apsu. In fact, there was no single definitive Mesopotamian cosmology, for different texts written in Babylon or Sumeria, perhaps many centuries apart, differ in their details. Even so, the Heaven, the Earth and the Underworld remain consistent general features, for this was a 'three-decker' Universe.

Water was of paramount importance in these cosmologies, and moved in an orderly fashion between the three layers – at least, when the gods were happy. The Egyptians saw the sky as a celestial river that reflected the Nile, while both the Babylonians and Jews spoke of great watery chambers existing above the visible firmament. The water could fall as rain, but then where did it go? Surely, it must drain away into the great underground waters of the Babylonian Apsu, or Egyptian Nun, out of which the world had first emerged at creation. For modern people living in northern Europe or north America, where the most common manifestation of water is in the form of rain or snow, it is not immediately obvious how the peoples of 2000 BC imagined the *sources* of water. Rain and snow, after all, are not common in the Near East, especially in Egypt. Yet if it does not pour out of the sky on a regular basis, where does the flowing surface water come from? With no knowledge of the rain cycle, or any clear geographical idea of the sources of the great rivers that flowed through their lands, it seemed to the Egyptians, Mesopotamians and Jews that their rivers probably bubbled up from out of the ground, just like the lesser springs and brooks. And their ultimate source was no doubt the great Apsu subterranean sea, that washed its way up through apertures in the thin pancake of the created world of men and women. Indeed, could this be what the composer of *Psalm* 42 was referring to when he spoke of 'One deep calleth another, because of the noise of the water pipes'? Subterranean water, indeed, was to remain a puzzle for millennia to come, as even Roman natural history writers like Lucretius and Pliny – who around the first centuries BC and AD were by then familiar with the arguments for a spherical Earth – struggled to explain its source. But in 1000 BC it was perfectly reasonable to think of the great rivers and big ocean masses such as the Mediterranean as connecting to the subterranean deeps by fissures and 'water pipes'.

And if this life-giving water somehow encircled the Earth, it became easy to see how the gods could cause flood or drought by playing with the taps. It is such playing with the taps, indeed, that lay at the heart of the ancient world's two great flood stories: those of Gilgamesh and of Noah. In both of them, either the gods, or God himself, chooses to punish a wicked world by inundation, and although the Gilgamesh story is probably older than the Hebrew story of Noah, and may have been a pagan antecedent to it, the same basic cosmological ingredients are present. According to the familiar *Genesis* version which describes Noah's ordeal, 'The windows of Heaven' are now opened, and a forty-day-and-night downpour descends. This floods the Earth even to its highest mountain tops, and only Gilgamesh or Noah and his family, in their respective narratives, are saved because they have built ships. Then, after a period of time (150 days for Noah) the waters are 'dried up from the Earth', and life begins anew. Exactly where so much water went to is not mentioned, but it was no doubt to its ordained place.

This, therefore, is how the three-decker Universe was seen as having been constructed, and how its regions worked together to explain natural phenomena. Yet such a cosmology remained plausible only so long as geographical knowledge was relatively limited, and especially while culture continued to be confined to land-locked peoples whose only acquaintance of the 'great salt sea' was restricted to occasional glimpses of the Mediterranean, the Red Sea or the Persian Gulf. Not until the age of the Phoenicians and Greeks, indeed, would increasing world travel fundamentally shake the foundations of this three-decker Universe; for it is my argument that it is *oceanic* experience that is primary in triggering a sense of a round Earth and encompassing sky. (This will be further explored in Chapter 4.)

The most tantalising references to great voyages of the archaic period, however, are those Egyptian accounts of the journeys to the land of Punt. Punt expeditions ran for more than 1000 years, and that of Queen Hatshepsut in 1493 BC placed her among the Pharaohs whose hieroglyphs and temple friezes recount the profits of these voyages, bringing as they did rare woods, liturgical substances, perfumes, gold, slaves, exotic animals and other luxury commodities into Egypt. The Punt voyages seemed to take a year or two for a round trip, and from items contained in the inventories, modern scholars have conjectured that the land of Punt was somewhere down the coast of East Africa – perhaps in the region of the Zambezi estuary. And could Punt also have been the land to which, around 1000 BC, King Solomon, at the very pinnacle of power of the new Hebrew Monarchy in Jerusalem, sent a fleet of ships on a three-year voyage down the Red Sea to the land of Ophir (probably off the Malabar coast of India) for gold, spices and precious substances? Solomon also traded in the opposite direction with the city of Tarshish (probably in Spain or north Africa) for exotic cargoes of 'gold and silver, ivory, and apes and peacocks', no doubt obtained from within Africa, as described in the First Book of *Kings* in the Old Testament.

Yet the most extraordinary of these long-distance voyages was that supposedly undertaken in the reign of Pharaoh Neccho around 600 BC, which set out to sail around 'Libya', or north Africa. An account of this voyage comes down to us not from direct Egyptian sources, but from the *Histories* of the fifth-century Greek tourist Herodotus. Included in the long narrative of his stay in Egypt (in which he became one of the first Europeans to leave descriptions of Egypt's already ancient monuments), Herodotus recounts the story of a voyage undertaken some centuries earlier. And while Herodotus speaks of Africa's being surrounded by water as though it were a generally known fact by his time of writing, he clearly has doubts about the factual accuracy of some of its physical details of which he was told.

Apparently, an expedition consisting of a fleet of Phoenicians in Neccho's employ sailed down the Red Sea and onwards down the coast of East Africa. But when the African coast finally began to turn in a westerly direction, the noonday Sun was found to stand on their *right* hand, or to the *north* of the sailors as they reached the tip of the African continent. This detail of the Sun standing to the north at noon was the fact which Herodotus admitted he found hard to believe, for the experience of people living in the northern hemisphere is that the Sun is always due south at midday, and is on the left hand when the observer is facing west. And ironically, it was this disbelief on Herodotus' part – hinging as it did on a piece of a natural phenomenon that could never be witnessed north of the equator – that shows that this extraordinary voyage must have happened. After a voyage of three years' duration, during which they had wintered ashore, the Phoenicians sailed through the Straits of Gibraltar and on to the Nile delta. So sailors in the employ of an Egyptian Pharaoh had been the first men to circumnavigate Africa.

Even so, there is no evidence that this, and the voyages to Punt, had the slightest effect on Egyptian astronomical or geographical thought. For it would need the different mentality of the Greeks, with their tendency for lateral thinking, to see geography as a science that inevitably worked in conjunction with astronomy, to fundamentally change the way in which the peoples of antiquity thought about the Earth and sky.

One of the ancient ways in which physical astronomy – as opposed to mythological cosmology – made its enduring impact upon culture was in the development of calendars for the regulation of civil and religious life. The earliest potentially calendrical physical struc- tures to survive, both in Egypt and in Britain, date from some time after 2600 BC; for the Great Pyramid of Giza, the Newgrange mound in Ireland, and Stonehenge were all constructed around this time.

The laying out of an architectural structure so that the Sun, or any other astronomical body, shines into particular parts of it at given times of the year, presumes a sustained and recorded observational familiarity with the sky; and while one can perhaps understand this taking place within the high civilisation of Egypt, it is quite breath- taking when thought of in the context of people who left no written records, for when dealing with structures like Newgrange or Stonehenge, the archaeological record is all that we have. And yet, Stonehenge possesses at least two clear astronomical alignments – those of the midsummer sunrise and of the midwinter sunset – while the 62-foot-long passage in Newgrange mound admits a thin ray of sunlight that would once have fallen upon the bones of its interred occupants at midwinter sunrise. Why these English, Irish and other Stone Age structures were constructed still remains a mystery, for their builders left no written evidence. But in the case of a monu- ment like Stonehenge, archaeology has shown that, far from being built all of a piece, this ancient circle was being modified and remodelled over a working life of nearly 2,000 years, and some physical features shifted, suggesting an emphasis away from solar to lunar calendrical criteria.

In spite of the complexity of monuments like Stonehenge and Newgrange, we still know virtually nothing for certain about the astronomical beliefs of their builders. This has not meant, however, that over the past three centuries there has not been a very great deal of speculation, most of which has been based on fortuitous alignments of the stones with particular astronomical events. In the *Choir Gaur, the Grand Orrery of the Ancient Druids* (1771), for instance, Dr John Smith produced the idea that Stonehenge was indeed the *Grand Orrery*, or astronomical calculating machine, of the Druids; while in the early twentieth century, the astronomer Sir Norman Lockyer believed that the astronomical secrets of its builders could be unlocked by careful surveys and mathematical analyses. And in the late twentieth century, Professor Alexander Thom took these astronomical arguments to even greater lengths. Nowadays, however, the desire to prove Stonehenge to have been some kind of 'megalithic observatory' seems to have passed.

While the midsummer solstice alignment at Stonehenge has been known for several centuries, it was really that remarkable eighteenth-century antiquary and eccentric, Dr William Stukeley, who launched the most tenacious of all Stonehenge myths: namely, that Stonehenge had once been a centre of Druidism. Yet we now know from carbon 14 dates that Stonehenge was probably already a ruin by the time that the shadowy ancient British Druids first appeared on the scene during the centuries immediately prior to the Roman invasion. (More will be said about modern Druidism in Chapter 12.)

While Stonehenge and other Neolithic monuments contain a variety of real and conjectural astronomical alignments, what cannot be disputed is the extraordinary accuracy with which the Great

Pyramid was laid out with regard to the four points of the compass. Indeed, ever since John Greaves made the first really accurate survey of the Great Pyramid in 1638, these coordinates have been noted. And in addition to the axial orientation, it also contains an inclined shaft on its north face which points exactly to the north celestial pole, although it is not clear what its original purpose might have been.

The Egyptians were probably the first humans on record to devise workable calendars and time-measuring techniques, both for purposes of religious liturgy and for civil administration. The earliest divisions of the Egyptian year seem to have derived from the waters of the all-important Nile, to produce the seasons of Flood, Subsidence of Flood, and Harvest. Yet while the sky mythologies of the Egyptians – with Ra, Nut, Thoth, and the deities discussed in Chapter 2 – were rich, their actual astronomical knowledge was limited to the functional. The Egyptians were, indeed, a naturally practical people who needed things that worked if Ma'at was to be maintained, Ra to be regenerated each morning in the eastern sky, and the dead to find their way to paradise.

From perhaps as early as 4500 BC, the Egyptians had noticed that just as the Nile was about to flood in early June, the star Peret Sepdet (Sirius) rose just before the Sun; and this 'helical rising of Sirius' nicely demonstrated the fitting together of divine and human affairs. And as Ra, in his nightly journey through the underworld, had to pass through twelve ritual portals, and recite the appropriate spells, the night came to be divided into twelve divisions. As Ra was believed to be entering a given portal, the position of a bright star just visible on the eastern horizon was noted. Its predecessors would now be climbing up the sky at approximately 15-degree intervals, so that one could reckon the number of portals through which Ra had yet to pass before he was reborn at dawn. Ra's passage through the portals was codified, especially at Ra's sacred city of Heliopolis in lower Egypt, in the *Book of Gates*, and here we find, in addition to appropriate incantations for each portal or Gate, references to the stars. In the tomb of Rameses VI at Luxor, the Keepers of the Twelve Gates are clearly seen depicted in human form.

This timing of Ra's nightly journey later came to be refined by the division of the celestial equator into thirty-six approximately equal spaces – eighteen for the day and eighteen for the night – denoted by a bright star, as different sections of the night sky became visible throughout the year as the Sun moved through the zodiac. Because of twilight, however, one could never see the stars across exactly half of the sky; therefore, six of these eighteen nocturnal star zones, or 'decans' – three at dawn and three at dusk – were of little practical value, and the day and night each came to be divided into twelve.

In the so-called Dendera Zodiac preserved on a temple ceiling at Dendera, we see the Egyptian gods in the sky in their full glory. Around the edge stand the thirty-six deities who represent the celestial 'decans', while other gods make up the other constellations. Although the Egyptians divided the sky into asterisms different from those finally canonised by Claudius Ptolemy in the second century AD, several bright Egyptian constellations have come down to us: the Plough (part of Ursa Major), for instance, was characterised as a hippopotamus by the Egyptians. But the Dendera Zodiac is not a star map in the modern sense: no actual stars appear on it – only a collection of stylised deities within a circular perimeter. For what concerned the Egyptians was not the physical accuracy with which the constellations could be measured in scientific terms, but how a bright star and its 'attendants' formed a recognisable marking point in the eastern sky against which to note the ritual passage of Ra's boat at its subterranean portals. The basic stimulus behind it almost certainly derived from the convenient helical rising of Sirius. But from it, we obtain our basic division of the night, and by extension, the day, into two sets of twelve hours.

Just as the Neolithic Britons at Stonehenge incorporated alignments to denote the northerly and southerly extremities of the Sun's rising and setting, so too did the Egyptians for Ra. At the summer solstice (around 16 July, as it would have fallen in 3500 BC), the Sun makes its most north-easterly rising and north-westerly setting, and describes the largest possible arc across the sky, to produce almost eighteen hours of daylight. And at the winter solstice (16 January in 3500 BC) the reverse takes place, with the rising and setting well south and tracing its shortest arc across the sky, with minimal daylight. Then, at what later Latin writers named the equinoxes (Latin for 'equal nights'), in April and October in 3500 BC, the Sun rises and sets on an east–west axis, resulting in twelve hours of light and twelve hours of darkness. The ancient Egyptians monitored these points in the solar year, marking each one with special rituals to maintain Ma'at. Central to the Egyptian calendar was the rebirth of Ra at around the winter solstice. According to religious belief, while Ra was the chief of the gods he needed not only to be refreshed nightly as he passed through the twelve subterranean portals on his Sun-boat, but also annually, and the reality of this rebirth was noticed in late January, as the days began to lengthen. And for this to happen, he had to pass through the body of his own *daughter* Nut, in that curious mix-up of parentage and generations so often encountered in mythologies.

When one looks at the Milky Way under a black unpolluted sky – as the ancient Egyptians would have done – one can see, with a little imagination, the long, thin starry body of Nut, with her head and her outstretched arms, as she arches over the world; and across the sky the Milky Way splits or forks to create the illusion of Nut's legs attached to her torso. Around 3500 BC, her mouth, as her head looked downwards, would have been in the region of sky that would become known to later generations as the constellation Gemini; and where her legs seemed to join on to her torso, near to the bright star Deneb in the constellation of Cygnus, would have been her genital organs.

Modern astronomers have calculated that on 19 April 3500 BC, over Egypt the Sun set at around 7.15 pm; and as it vanished below the horizon, that region of the Milky Way where Nut's mouth was believed to be would have become visible in the darkening sky, giving rise to the sense that she had swallowed Ra. The Sun god would then pass through Nut's body for the next 272 days – nine months – and would be reborn at the winter solstice on 16 January. At this time of year, Nut's outstretched legs would have risen before the Sun, so that Ra would have been apparently reborn from Nut's starry body to begin the world anew. Astronomical observation, mythology, and the eternal cycles of the Egyptian state, all fused into one great truth.

Today, however, the Sun does not quite move in the same way. Because of the precession of the equinoxes – caused by a tiny annual discrepancy between the motions of the Sun and the Earth (described in Chapter 5), and which cycles over a period of 25,800 years – the Sun's position relative to the stars is now slightly different from what it was 5,500 years ago. This is why, for instance, the solstices and equinoxes now fall in June and December rather than in January and July as they did in Egyptian times, and why the Sun does not quite relate to particular regions of the Milky Way as it did when it was thought to constitute Nut's celestial body.

But it was from a careful counting of the days necessary for Ra to return to the same spot in the sky, or rise in accordance with the same distant horizon marker, that the Egyptians were able to calculate the length of the year. At first, it seems to have been reckoned as 360 days in length, although as far back as 3000 BC at least, the extra five days – won by Thoth for Nut to give birth to her children – had extended it to 365 days. These days were ordered into twelve months of thirty days each, with the five days added to the end of the year.

But as the Egyptians, like many other cultures, realised, the convenient lunar month – so useful for marking the passing days from the changing phases – did not synchronise with the solar year, for twelve lunar months produced only 354 days. Far back in their cultural history, therefore, the Egyptians had devised a series of empirical formulae whereby they tried periodically to intercalate their calendars and make the natural astronomical points of the year – such as Ra's rebirth at the winter solstice – agree with the religious and civil calendar.

In the way that one of the principal driving forces behind the Egyptian calendar – especially that used for religious liturgical purposes – was the maintenance of Ma'at on Earth and in heaven, so the Mesopotamian calendar was largely concerned with providing a basis for divination. Celestial divination, just like animal liver divination, was seen not as a hard and fast fate, but as a forewarning of what the gods could do unless steps were taken to placate them. This meant, therefore, that official or state astrology was to be a major impetus behind Mesopotamian astronomy.

Mesopotamian sky lore probably began in Chaldaea and Sumeria around or earlier than 3000 BC, when some of the constellations still familiar to us today – such as Aries the Ram and Taurus the Bull – first made their appearance. In the late nineteenth century, in fact, the English astronomer Richard Anthony Proctor tried to calculate the age of some of the Mesopotamian constellations that have survived through the Greeks and come down to us today. Proctor did this by using the annual rate of the precession of the equinoxes, which by his time in the 1870s and 1880s was known as accurately as we know it nowadays. He worked on the assumption that when the constellations below the celestial equator – such as Argo the Ship – were first devised, they probably moved in upright positions across the horizon, as they came due south, as seen by people living about 30 degrees above the equator, such as the ancient Sumerians. But today, many millennia later, they do not seem to stand straight up, and some of their stars seem less obviously placed than the original constellations would appear to indicate. From these, and from other criteria of the precession of the equinoxes, Proctor calculated that the southern constellations would, indeed, have stood upright on the southern horizon around 2170 BC. And in the absence of any early documents of this period specifying their names, Proctor's estimate is probably as good as we can hope for.

But regular sky watching and recording in Babylon really did not start until after 2000 BC. Like the Egyptians, the Babylonians were also faced with producing workable lunar and solar calendars for the business of ordinary life, as well as state prognostication. Aware as they were of the discrepancies between the lunar and solar years, they followed the practice of intercalating extra months as necessary, as a way of making their early 360-day calendar, consisting of twelve months of thirty days each, square with reality. In the eighteenth century BC, King Hammurabi was already ordering the insertion of these intercalary months, and subsequent cuneiform tablets (the wedge-shaped script impressed into clay in which the Mesopotamian languages were written) indicate that this became a standard calendrical practice for centuries afterwards.

The Babylonians viewed the sky differently from the way in which the Egyptians viewed it. Perhaps the greater instability of their environment, and the more slave-like relationship between humans and their gods, made divination and divine forewarnings more important to the Mesopotamians. Whatever the case, the result was an extreme sensitivity to the relationship between Earth and the sky, and to how their gods in the sky displayed patterns of behaviour of which it was salutary for humans to take note. For it was their concern with celestial divination which supplied the impetus for systematic astronomy and mathematical analysis. One might plausibly argue, therefore, that it was the religious beliefs of the Babylonians in particular that saw the birth of observational and mathematical astronomy.

The earliest recorded astronomical observations of the Babylonians have to do with the motions of the Moon, the dates and times of eclipses, and the extreme 'elongations' of Venus east or west of the Sun during the morning and evening. Because Venus can never be more than 45 degrees away from the Sun-god Shamash, she is only ever seen just before or just after sunset, and can vary considerably in brightness depending upon her position, making the morning and evening star a significant astronomical body, and worthy of study. Tablets dating from the reign of Ammisaduqa in the eighteenth century BC record some of the earliest observations of this kind.

Several Babylonian manuals of astronomy have come down to us. The one known as *Mul. Apin.* lists some sixty individual stars and constellations, and speaks of up to eighteen constellations spread around the zodiac. The zodiac was of particular importance, being that band of sky through which the Sun passes over the course of the year. And while one can never actually see the Sun in a particular constellation because the stars are not visible in the daytime, one can come to establish the Sun's annual course if one is sufficiently familiar with the night sky. For the shapes and mutual relationships of the stars and constellations never change, and for a given place on the Earth's surface they always rise and set against the same spots on the horizon, and reach the same elevation in the south. What does change, of course, are the times of night at which they rise and set, for this depends on the seasons. At 10 pm on an October night, for instance, the ancient constellation of Taurus, with its bright star Aldebaran, is rising in the eastern sky, whereas at the same time in January it is high up in the south, and by 10 pm in mid-March it is setting in the west.

It eventually came to be realised, therefore, that if the Sun moves around the sky in 360 or 365 days, one can tell which stars it must be amongst, when they are rendered invisible by its light, from those that appear in the night sky after it has set. For all one has to do is draw a line across the circle. Yet what this realisation requires is what might be called a leap of abstract thinking, based on numbers and proportions; but such a process of abstraction is not possible until one has developed a concept of numerical and physical regularity in nature.

In the same way that the Egyptians had divided the sky into thirty-six decanal zones, each denoted by a bright star and its dimmer consorts, so the later Babylonian zodiac had thirty-one bright stars chosen to form roughly equidistant sky zones to act as marker points against which to track the motions of the planets. The zodiacal band became the region of sky to excite the first astronomical interest of the ancients because not only the Sun, but also the Moon and the rest of the planets, move within it. Though there were large and conspicuous constellations that were not part of the zodiac – such as the Plough (part of the Great Bear, or Egyptian hippopotamus) in the north polar skies, and brilliant Orion with his three-star belt and attendant Sirius in the south – these constellations were never seen to be occupied by the planetary divinities. Indeed, the Moon, and the planets Mercury, Venus, Mars, Jupiter and Saturn (in their familiar Roman names), can never wander very far beyond what we now know is a flat plane that passes through the Sun's equator and extends across the solar system. And while the ancient Babylonians would not have known about the gravitational physics which causes the phenomenon, they would, nonetheless, have known its effect: namely, that the Moon and planets are confined to a narrow band of sky which is the same as the Sun's path among the stars.

For this reason, therefore, it was necessary, early in Babylonian culture (probably in the second millennium BC), to establish the above-mentioned marker points of unchanging fixed stars along this zodiacal band, against which to record the planetary wanderings. For it is these observations, written into wet clay with a wedge-shaped stylus, fixing the positions of the Moon and planets with regard to the Sun for specific dates, that constitute the earliest 'scientific' astronomical records: scientific in the respect that they record not only the moral or legendary attributes of the gods in the sky, but also their geometrical relationships with the Sun and with each other.

By the fifth century BC at least (and probably for a long time before it), this band had finally come to be divided into the familiar twelve signs into which we still divide the zodiac, or ecliptic. Indeed, a boundary stone of *c.*1100 BC, preserved in the British Museum, shows engraved images not only of the Sun, Moon and Venus, but also of a scorpion and a lion, which *might* be early depictions of those constellations later named Scorpius and Leo. It was also this zodiacal band, divided into its twelve signs, through each of which the Sun passed for a thirty-day period, which no doubt gave subsequent astronomical science that priceless computational tool: the 360-degree circle. Indeed, without this elegantly divisible number – with its 30, 90, 180, and other proportional units – it is difficult to imagine how either mathematical astronomy or coordinate geometry could ever have come into being. For while the natural properties of the circle – such as the capacity to exactly accommodate a hexagon formed of six radii – are independent of any man-made units of measurement, the ancient liturgical year of 360 days through which the Sun god was believed to pass, at the rate of one digit per day, provided an ideal number for the division of circles. It is difficult, indeed, to imagine a coordinate geometry based on the division of the ecliptic into 365¼ units!

It was the development of mathematics which revolutionised Babylonian astronomy, and elevated it above a patient listing of eclipses and omens, to provide it with a unique and enduring power. This was, moreover, a new concept of reckoning that went well beyond the type of counting that was necessary for parcelling out land or counting royal tributes. It was a conception of numbers or properties that could be manipulated abstractly to produce far-reaching conclusions; and vital to this process was the anonymous invention of variable place numbers rather than symbols for absolute quantities (for example, 50 or 100) such as the Egyptians used.

Perhaps because it forms a natural one-sixth division of the 360-digit solar year, and corresponds to the radius of a circle divided into its circumference, the Babylonians developed a sexagesimal (base 60) system of reckoning. Place numbers made it possible to devise symbols for single digits and different ones for larger units, such as 10 or 60, all of which could be cut by a stylus into wet clay. Numbers such as 8 or 33 could therefore be built up from the appropriate characters. Clay tablets inscribed with mathematical schemes depicting numbers up to 59, with compounds thereafter, still survive from as far back as 1700 BC. Some of these tablets have even led scholars to believe that as far back as the second millennium BC Babylonian mathematicians could calculate simple square roots of numbers or solve problems in simple trigonometry – which would have been impossible, and probably irrelevant, in Egypt. Such flexible numerical systems enabled the development of the first intellectual mathematics in Babylon, which was in turn to provide a vital foundation for a scientific approach to astronomy. Yet all of this was done for politico-religious purposes, as a way of obtaining deeper insight into the omens and forewarnings with which the celestial deities threatened the state.

Although Babylonian astronomical 'Diaries' or records date from before 1000 BC, it was not really until the time of King Nabonassar, around 740 BC, that consistent runs of observations were recorded, written down in chronological columns. Even Ptolemy himself, when compiling his *Almagest* around AD 150, confessed that reliable runs of Babylonian observations were not available before Nabonassar. But after this time, and certainly by 600 BC, Babylonian astronomy was developing systems of analysis and mathematical explication of phenomena of astonishing sophistication.

One of the most significant of these achievements is the discovery of the mathematical mechanism underlying the eclipse cycle. While both the Sun and Moon move within the band of the zodiac, the Moon's orbit within that band is so complex as to defy any simple explanation. We now know that this complex lunar orbit is a product of what Sir Isaac Newton was later to call the 'three bodies problem', as the Sun, Earth and Moon all exert mutual gravitational influences that are constantly changing as the bodies move with respect to each other. And as the Moon is by far the least massive of the three gravitational forces, it describes a frightfully elaborate dance within the zodiac over any period of years. The Babylonians knew nothing about gravity, but they dated and timed records of eclipses, as well as of lunar 'syzygies' – times each month when the Moon, the Sun and the Earth all lie on the same straight axial line. At such times an eclipse can occur, although the Moon more often passes either above or below the Sun, or else the Earth's shadow likewise misses the Moon.

As part, no doubt, of trying to refine their knowledge of the inconveniently asynchronous solar and lunar years for purposes of intercalation, the Babylonians realised, by the fifth century BC, that a run of 235 lunar months came very close to 19 solar years. This provided astronomy with a predictive power that it had previously lacked. On the one hand, this 19-year cycle (later called the 'Metonic cycle', after the Greek who brought it to a wider audience) was able to provide a mechanism for the reliable prediction of eclipses. From the records of lunar syzygies dating back to 800 BC and earlier, Babylonian astronomers were able to notice the patterns in which eclipses occurred. However, far more important in the wider scale of scientific history were the components which Babylonian astronomers had brought together by the sixth century BC. These included the concept of making and keeping long runs of precisely dated astronomical records, such as the date, time and degree of obscuration during a particular eclipse. Another vital component was their realisation that these records could be made to yield up regular sequences in which events such as eclipses of the Sun and Moon took place, to initiate the concept of scientific 'data' from which reliable future predictions could be made. Their development of the base 60 and place numbers also provided them with flexible computational tools with which to handle observational data; for to the Babylonians, simple counting had matured into a conceptual system of *mathematics*. As the Greeks would soon discover, this opened up a whole realm of ideas that made it possible to tackle problems in coordinate geometry or statistics in which shapes and numbers could be used creatively with precise systems of rules to discover things hitherto beyond understanding.

One example of this conceptual mathematics in action was the development of 'zigzag' functions, aimed at finding a solution to a type of problem that in a much more complex form was later to puzzle Sir Isaac Newton, and even had to be solved by NASA in the late twentieth century when developing 'slingshot' techniques for using planetary gravity to project spacecraft such as Voyager into the depths of the solar system. The problem, in brief, is as follows. How does one express a mathematical relationship when the object under study – such as the Sun or Moon – does not move at a constant speed, but at a variable speed? The zigzag computational technique invented by the late Babylonians, and used in *Enuma Anu Enlil*, introduced the concept of a *mean* speed being ascribed to an accelerating and decelerating body, for the Moon over the course of the month and the Sun over the year move at variable speeds with respect to each other. In consequence of these techniques, the post-500 BC Babylonians were able to develop the first theoretical 'models' for the motions of the Sun, Moon and planets, with particular attention being paid to the Moon, which was of such calendrical importance.

All of this was not for reasons of intellectual curiosity, but in the service of the state and religion. By being able to establish the patterns in which the celestial deities moved, then perhaps one could heed their warnings more effectively. In Babylonian thought, human beings were the slaves or underlings of the gods, and lacked that special moral and design relationship with their gods that the Jews had with Yahweh in subsequent intellectual history; but the slaves were nonetheless clever. Indeed, one might think of them as rather like the competent, situation-saving slaves that were to become almost standard characters in later Greek and Roman comedies – characters who were really cleverer than their masters, and were adroit at preventing family disasters, such as the fictitious slave Tranio in Titus Maccius Plautus' (*d.*184 BC) comedy *Mostellaria* (*The Ghost*). (Tranio, in many ways, was similar to Frankie Howerd's Roman slave Lurkio in the 1970s television and film comedy *Up Pompeii.*) And perhaps, like these clever human slaves, the Babylonian astronomers and astrologers – with their mathematical techniques for being able to tell what the gods in the sky were to do next – could prevent disasters on Earth by interpreting their forewarnings and recommending ritual countermeasures.

But before one can use sophisticated mathematical techniques to reliably predict celestial phenomena, one must have that bedrock of data, such as that recorded in the Babylonian tablet 'Diaries', upon which to employ them. The Babylonians had realised that, no matter what the status of any given mythical narrative, the heavens moved at their own speed, and needed to be observed and monitored if the predictions were to be of proper use as warnings.

How, therefore, did the Babylonians actually observe the heavens? There is no evidence that they used graduated instruments divided into parts of circles. Instead, a planet's motion through the zodiac was noted from its contiguity at any one time to preselected evenly-spaced stars, such as those that formed the nightly 'decans'. Planets moved through celestial hours, or decanal zones, rather than through 'angles'. Even so, it was still possible for an experienced observer to be remarkably accurate in his estimates, especially if finger-breadths held up at arm's length were used to aid the eye. A man's little finger held out at arm's length, for instance, subtends an angle of about 1 degree of arc – approximately twice the diameter of the Sun and of the full Moon (which both subtend the same angle), or $\frac{1}{180}$ of the zodiac. I have found from personal experience that with practice one can trace quite accurately the motions of bright planets, such as Jupiter, Venus, and, of course, the Moon, by estimating their changing nightly positions against familiar zodiacal stars such as

Regulus (in Leo) or Aldebaran (in Taurus). It becomes even easier if one adds the instrumental refinement of up-held fingers! One can also use one astronomical body to find the position of another. When the Sun's light is dimmed by the Moon during a total solar eclipse, for instance, it becomes possible to see bright stars in the sky for a few minutes, thereby confirming the Sun's position in a particular part of the zodiac. Likewise, a total eclipse of the Moon, or even an ordinary full Moon, occurring as it does when the Moon is exactly opposite the Sun when seen from Earth, can be useful for determining the zodiacal place of the Sun. If, for example, a lunar eclipse occurs when the Moon is within a finger's breadth of the bright star Aldebaran in Taurus, then it is easy to deduce – by casting an imaginary straight line across the circle of the ecliptic – that the Sun must be very close to Antares, in Scorpius. This celestial geometry, moreover, is quite independent of theories of a moving or stationary Earth. And the Moon's phases also constitute a natural clock and position marker in their own right. For people living under clear skies (clearer than those over England), where astronomical bodies can be relied upon to appear at sunset, and are the familiar seasonal lights, it is possible to trace their movements with remarkable accuracy on a nightly basis.

On the other hand, it is also known that there were some astronomical instruments familiar to the peoples of the Near East. Large buildings such as temples, and the pyramids, were laid out in accordance with precise coordinates. The Egyptian *merket*, which consisted of a vertical plumb line and distant viewing slit cut into a piece of wood or reed, could be used for surveying long straight lines that could act as horizon marker points for the rising and setting of astronomical bodies. Similarly, several Egyptian sundials survive from tombs, and an extant inscription from the time of Pharaoh Seti I (*c.*1300 BC) shows how to use such a dial. Egyptian dials tended to be T-shaped, portable, and made of wood. The short arm of the T was raised up, and when the instrument was pointed at the Sun on a level

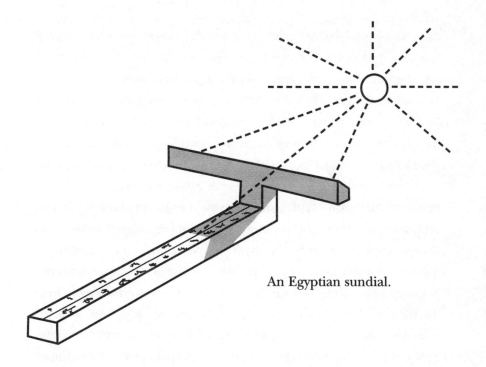

An Egyptian sundial.

surface, it cast a shadow on the long arm. This long arm would carry graduations which made it possible to read off four hours before noon and four after noon, although the shadows would have been too long for it to have been effective for the hours nearest to sunrise and sunset. One would also have needed different dials, or sets of scales, for different seasons, to allow for the high and low Suns of summer and winter.

One of the most ingenious ancient instruments – also Egyptian – was the water clock. These were the earliest artificial timekeepers, and in the Oriental Institute in Chicago there is a beautifully preserved specimen from the New Kingdom (1567–1320 BC). It is shaped like a large flower-pot, its conical sides assisting in maintaining a fairly even water pressure as the vessel empties through a tiny exit hole. Inside the vessel are monthly scales, so that while the water empties at a fixed rate, the device can still be used in different seasons of the year by reading the proportionate scale for that season. And very appropriately, it carries the effigy of a baboon, which was one of the guises of Thoth, the Egyptian god of time and measurement.

Such water clocks could serve a variety of purposes. They could mark the days when the sky was cloudy, to ensure the continued performance of the appropriate rituals; and they could also be used to measure the periods between the risings of stars, or their passages across the meridian, for there still exist texts describing the observation of meridian transits as well as risings for time-measuring purposes. The best evidence of astronomical instruments comes from Egypt, probably because of their survival in tomb collections; but even so, the archaic cultures were not instrumental in their approach to astronomy. While the Egyptians and Babylonians had divided the day and night into twelve hourly periods, devised early ritual calendars that lay the foundation for the 360-degree circle, learned how to relate the wandering planetary deities to proportionate places within that circle, and even to predict their future movements, along with inventing both the concepts and techniques of mathematical analysis, their cultures were what might be called cumulative rather than investigative in their use of knowledge. The intellectual and spiritual energies of their societies were directed towards maintaining religious harmonies and social order, rather than towards any abstract investigation of nature or the world of ideas for their own sake. Their cosmologies remained concerned with flat planes and the movement of water, while the whole of creation was seen as both the product and the plaything of temperamental deities.

In the sixth century BC, however, this three-millennia-long world view was to change in the north-eastern Mediterranean, with that wind that blew across Attica and ushered in a new way of viewing the cosmos and a new direction in the history of human thought.

FOUR

The wind that blew across Attica

For several centuries, scholars have tried to define what led the Greeks to fundamentally change the direction of human thought, and to determine why it was that, for at least 2,500 years, the great civilisations of the Near East had, in spite of political turbulence, maintained cultures that seemed eternal in their basic assumptions, whereas in Greece, affairs were to take on such a different shape, to generate a society which we can recognise today as the ancestor of the modern world.

For it was in Greece, mainly from the sixth century BC, that a number of recognisably 'modern' concepts first emerged. These included political and economic freedom, assumed rights to self-government, and even a degree of choice in religious belief. They also included ideas in physical science and medicine, for the natural world was held to possess an independent intellectual integrity that was quite separate from the whims of squabbling gods. And art, further-more, came to be seen both as possessing an intellectual and spiritual status in its own right, and as public, to be enjoyed not only by the citizens who paid for it, but also by society at large.

But why Greeks were so pivotal in world history was, perhaps most of all, their recognition not only of the centrality of reason, but also of the necessity for personal freedom, in which reason could be exercised and truth revealed. To the Greeks, the rational intellect was the distinguishing hallmark of humanity, for it links us to and gives us insights into eternal truths, and leads us on to a perception of the divine.

The burgeoning mercantile wealth of the Greek world, based as it was on overseas trade and the profits of shipping, also created a new social class. Greece was the first 'middle-class' society, made up largely of people who were neither aristocratic war lords nor peasants, but who had wealth and leisure, and chose to use them in developing the sophisticated urban life of the independent 'polis' or city state. These citizens not only spent their wealth in creating a pleasing physical environment, but also in giving high status to artists, architects, actors and musicians, as well as initiating the modern professional groups of lawyers, doctors and academics.

From this world, too, emerged the concept of individualism. The names of specific ancient Egyptians or Babylonians that have come down to us have invariably done so because the people in question represented particular types within their societies: rulers, good servants, lawgivers, or scribes. But they rarely ever emerge as three-dimensional people with tastes and preferences unique to themselves. And while the Jewish Old Testament contains narratives of countless named people, their lives likewise are invariably recorded because they embody wisdom, folly, prophecy, or obedience or disobedience to God.

The Greeks. however, spring on to the scene as real people. The autobiographical travel narratives of Herodotus from the fifth century BC; the eating, drinking, arguing real-life and named characters who populate Plato's *Dialogues*; or the human types satirised in the plays of Aristophanes – all indicate a fascination with human peculiarity.

This individualism was a natural and encouraged component of Greek society, manifesting itself not only in military heroics, but also in competitive sports, ingenious legal pleading, poetry, oratory, philosophy and science. For while no-one knows who wrote the Babylonian *Enuma Elish* or the Egyptian *Book of Gates*, few Greeks, it seems, cared to put pen to paper without appending a name – be it the writing of a geometrical treatise by Euclid, or a poem by Sappho. In short, it was the Greeks who invented intellectual personality, and formally attributed new ideas to the men and women who conceived them. Unlike the monolithic, eternal empires of Babylon and Egypt, the Greeks craved novelty. The numerous different 'schools' of philosophers that emerged – the Pythagoreans, Eleatics, Sophists, Stoics, Platonists, Aristotelians, and so on – testify to this fondness for intellectual searching, and when St Paul (himself a Greek Jew) first preached Christianity in Athens, probably around AD 50, he capitalised on the Greek love of new notions by proclaiming Christ as 'The Unknown God', to whom he had seen an altar dedicated, 'For all the Athenians... spent their time in nothing else but either to tell, or to hear, some new thing' (*Acts*, 17:21).

And what factors might have conduced to these extraordinary innovations in Greece? Though historians admit that it is not possible to provide all-embracing explanations for such prodigious cultural developments, one can at least point to some important preconditions that might have helped to prepare the way. One essential ingredient was political freedom. The Greek city states, with their citizen 'hoplite' armies and concern with participatory government, were no doubt aided by the country's topography. Its islands, mountain valleys, and often jagged terrain, rendered a centralised political system impossible; and around 350 BC, indeed, Aristotle studied the constitutions of 158 city states, and acknowledged that this number was only a fraction of the autonomous political units that existed across the Greek world. This was, of course, a terrain in stark contrast to the flat lands of Babylon and Egypt, with their excellent river communications and resources for social control.

One might also argue that the very presence of the sea around Greece made its peoples natural sailors from the remotest times. This maritime culture not only gave rise to traditions of commercial activity and far-flung trade, but also to a situation in which men were accustomed to pitting themselves against a dangerous environment of storms, treacherous seas, and the daily risk of shipwreck. It was, in short, a culture in which courageous individual initiatives were both expected and applauded, for a man who is accustomed to risking his life to bring home his precious cargo to port to be sold for a profit is less likely to submit to arbitrary authority than is a frightened peasant. It is hardly surprising, therefore, that when the Persians tried, and failed, to conquer Greece in 480 BC, they could not understand why the Greeks fought so fiercely to defend these peculiar things called freedom and independence.

This maritime culture, one might suggest, also created a geographical awareness in the Greeks that would have been less likely to arise in a land-locked culture. For if one sails the oceans, certain things sink into one's broader awareness – such as the realisation that when sailing over the horizon, a ship's hull disappears first, and the flag at its mast-head disappears last. And should one then climb a cliff, or the mast of one's own ship, one can see the departing ship's hull reappear above the horizon, only to slip away again. And why is the same solar eclipse, visible at noon in Athens, seen at 9.00 am in Spain and at 1.00 pm in Egypt, when respective sundial timings are collated? Why, moreover, if an eclipse is a lunar eclipse in which the Sun sets on one horizon as the full, shaded Moon rises on the other, is the shadow that falls across the Moon always curved? It is not for nothing that the first coherent arguments for a spherical Earth, rather than the traditional notion of the Earth being a flat plate, came from the Greeks.

Prosperity and political freedom also brought leisure. Men now had the spare time to compete with each other in sports, watch plays, discuss cosmology, and explore profound subjects such as the nature of truth and the eternity of geometry. It was this culture which gave employment to professional teachers and thinkers such as Socrates, and invented the discipline of philosophy.

Early Greece – from about 1200 BC onwards – also had a rich bardic culture, in which the *Iliad* and *Odyssey* of Homer told the Greeks about their supposed heroic origins, as well as narrating fabulous adventures such as those of Odysseus on his way home after the Trojan War. By the eighth century BC, however, Hesiod, in his *Works and Days*, was turning his attention to less exalted figures such as farmers, who used the rising and setting of the stars to assist them in framing a calendar. In September, for instance, 'when Orion and Sirius are come to mid-heaven, and rosy-fingered Dawn sees Arcturus', recommends Hesiod, 'then cut off the grape clusters... and bring them home' (Loeb edn., p 49). Likewise, at the end of October, 'when the Pleiades and Hyades and strong Orion begin to set, then remember to plough in season' (*Ibid.*, p.49). Hesiod's farmers, moreover, were not serfs bound to the soil, but free men who were (presumably) willing to drop their spades and take up their spears to defend their ground.

It is also in Hesiod's *Theogony* (on the birth of the gods) that one encounters the first coherent Greek religious writings, as deities often mentioned in Homeric or pre-Homeric traditions came to be presented to the world in a series of Greek creation myths. According to Hesiod, the world came into being when the great void of the cosmos and the Earth force – personified in the deities Chaos and Gaia – were brought together by Eros, and the procreation of the gods began. Kronos, who had castrated Uranus, was overthrown by Zeus, along with the twelve Titans and Giants. Zeus then emerged as the king of the gods, though in no way was he their creator. Then were created mortal men and women, who went on to populate the world.

These, however, were not the myths that had such a powerful influence in forging Greek perceptions of origins and purposes, as had been the creation stories of the Near East. Instead, the Greeks searched for non-personal, enduring principles as a way of providing coherence. Indeed, it probably tells us something about the Greek frame of mind that the only culture in the ancient world to see creation not in terms of personified actions so much as indicating the presence of an intellectual principle, should also have been so assiduous in committing to record the names of its philosophers.

To the Greeks, this enduring intellectual principle was the *logos* ('word', 'speech', or 'reason'), which was first expounded by Heracleitus in the sixth century BC. Yet this *logos* was not the creator so much as an active principle of order that suffused everything, and was the opposite of chaos. It was Anaxagoras (*c*.500–428 BC) who developed the *logos* into one of the most influential concepts in Greek thought, and defined a principle that was above the world and greater than the physical creation, and without which the world would fall apart. This concept was the *nous* ('mind' or 'intelligence').

These ideas, indeed, would have a powerful formative influence on subsequent Western thought. The *logos* concept would later be used as an intermediary between God and human beings by Philo of Alexandria (*c*.30 BC–AD 45) in his synthesis of Greek and Jewish thought, while the *nous* would perhaps find its definitive expression in the writings of the Neo-Platonist Plotinus in the third century AD. The idea of a universal and enduring Truth that suffused all things of the intellect was also to constitute one of the great themes running through Plato's *Dialogues* from about 390 BC.

Plato's *Dialogues* are one of the great milestones in the history of human thought, for they embrace a number of universal elements – profound theological insight, discussion about what comprises the civilised life, the respective values of action and leisure, and the nature and definition of truth and love. They are, moreover, set – as their names indicate – in the social context of dinner parties and debates, in which real historical personalities such as Socrates, Protagoras, Crito, and others, voice opinions which are subject to cut-and-thrust argument, which Socrates invariably wins! Here are real men, reclining around tables overlooking the sea, eating and drinking on summer's afternoons, as they explore the nature of the human mind – all of which produces in Plato's *Dialogues* a human immediacy that is a world removed from the divine struggles depicted in Egyptian and Babylonian literature.

Running through the Platonic *Dialogues* is a body of ideas about creation, change and eternity which was the most sophisticated statement on those subjects to come out of the ancient world, and which was to have the profoundest consequences for human thought. And because Plato's discussion of creation and order was within the context of a single, unified design principle, it was useful not only to later Jewish philosophers such as Philo, but also to St Augustine, Origen, and subsequent Christian theologians who saw Plato's thought as prefiguring insights that would be brought to full realisation in Christ.

Greek religion possessed a pluralism which made it unique, for while there were stories of creation and of Kronos and the other deities related by Hesiod and others, and while Zeus could, at least in theory, take the form of a swan to seduce the nymph Leda, these were not tales of ever-present or potential terrors, as were the myths of Babylon or Egypt. Indeed, from the sixth century BC onwards, some Greek philosophers were even so bold as to raise serious doubts about the very existence of the traditional gods. Xenophanes (570–480 BC), for instance, had argued that only one Supreme Being existed, while over a century later Xenocrates – who was head of the Athenian Academy between 339 and 314 BC – said that there were only the astronomical deities of the Sun, Moon, Mercury, Venus, Mars, Jupiter, Saturn, and the stars. And Cleanthes (331–232 BC) also doubted if there were any individual gods in addition to the one Supreme Being. Yet Xenophanes' and Cleanthes' Being was not the same sort of deity as the Jewish Yahweh, for he 'resembles mortals neither in form nor in thought', and man is not made in his image. Indeed, this Supreme Being, who was self-created and who never moves and never changes, is really a principle of pantheism, in which one Being suffuses all things. Even so, Xenophanes was perhaps the first great philosophical theologian in history – the first thinker to penetrate beyond the simplistic attribution of human characteristics to the gods, and to instead search for enduring intellectual and spiritual principles.

But what made such radical theological innovations possible for the Greeks – innovations, moreover, which from the very beginning are ascribable to named, historical personalities rather than to legends? Indeed, it is difficult to separate these intellectual and spiritual freedoms from the wider political and economic freedoms discussed above, or from that fascination with human excellence and individuality which were natural outgrowths from them. Yes, a person might offer a sacrifice, or go to Delphi to consult the famous oracle; yet the advice which one received was not rigidly prescriptive, but could be acted upon or interpreted as one saw fit.

As with the Egyptians and Babylonians, one of the central concerns of Greek religious thought was the abiding need to explain creation, order and change. Where the Greeks were so innovative in this quest, however, was in their inquiry into the very nature of perception itself; for whilst the peoples of the Near East had taken the human senses at face value and had devised their myths accordingly, the Greeks began to wrestle with the problems of *epistemology* itself (*episteme* – knowledge; *logos* – rational understanding, science). The Greeks were, in fact, to invent the very techniques through which rational inquiry was to be explored, be it in physics, medicine, psychology, geometry, logic, or theology; and central to this inquiry was the need to explore behind appearances, to first recognise the pitfalls to which the human mind was naturally prone, and then to pass on to the universal truths that lay beyond.

Here, indeed, lies the supreme contribution of the Greeks to civilisation: their recognition that the ordered and disciplined mind really could penetrate the fogs of myth, opinion, and vague specula-tion, to gain a clear apprehension of what was eternally true and divine. Greek philosophy, therefore, was profoundly religious in its quest, and all the arts and sciences which the Greeks formulated were not thought of as human inventions but as those revelations of the eternal that could be uncovered only by reason.

One of the driving forces which motivated Greek philosophy was the true character of *episteme*. Was it related to the five senses, or was it purely intellectual? Could we learn about the world and the Universe by careful looking at it and measuring it, or could we only perceive absolute truth by passing beyond sensory knowledge into the realm of *ideas*, of *logos*, or of *nous*?

It was the school of Greek philosophers which had formed around Xenophanes, and was to take its name from its eventual place of settlement – Elea, in Italy – which first discussed these matters. The Eleatics considered that the five senses only led one astray, and that one could never reach the eternal through their use. Physical percep-tions, after all, were liable to vary, as wind, storm, heat, cold and colours were often perceived differently by different people. It was, indeed, Xenophanes' disciple Parmenides (*c.*515–449 BC) who really formulated what was to be the classic Eleatic position, in his poem *On Nature*. In its Prologue, Parmenides spoke of the search for truth as a road which divides. One branch – that of the senses – leads to night, because the often conflicting nature of sensory knowledge can only ultimately take one to darkness. The branch for the true philosopher to follow, however, was the pathway of pure intellect, or self-evident reason, which would lead to the full light of day. Parmenides, therefore, became one of the founding fathers of that philosophical school called rationalism, or idealism: ideal, indeed, because only *ideas*, and not objects, possess ultimate reality. According to Parmenides, therefore, change is really an illusion. The things that change – such as the seasons, or men's bodies – are only shadows of the night. Ideas, on the other hand, never change, because they are part of the divine.

However, Heracleitus – who lived in Ephesus around 500 BC – argued that the physical world itself enshrined profound truths, and was worthy of study in its own right. For while the physical world was full of change, that very force of change possessed in itself an eternal truth. What fascinated Heracleitus, in fact, was the creative tension between change and underlying permanence, and he cited several examples. We as people, for instance, are aware of changes within our own bodies, yet in spite of it all, our own sense of self does not change. Heracleitus also used the analogy of fire, saying that while flames constantly dart, move, and die down, fire itself is nonetheless eternal. This led him to argue that fire is the truly eternal principle behind all being: one of the earliest conscious expressions of an 'element', in fact. And perhaps Heracleitus' most famous analogy in exploring the paradoxical relationship between change and permanence was that of flowing water. One can never step into the same river twice, for each time one steps into the river, a different body of water flows over one's feet; yet this does not in any way undermine the concept of a river as a permanent thing. Individual units of change, in other words, do not invalidate the greater truths that lie behind them.

All things, therefore, exist in a state of flux. This flux, moreover, should not be seen as chaos, for it is the natural expression of the supreme truth that lies beyond the creative tension of paradox. This is how Heracleitus conceived of the *logos*: as that power of potential which produces the actual, and which links all things together.

But what was the very substance out of which all things were made? Once again, it was the Greeks who first asked this question for, one might say, it would have been meaningless to a Babylonian or Egyptian. To them, flesh, metal, wood and water existed as obvious facts of life, because the gods had brought them into being when the world emerged out of nothingness. And while it is true that all the Near-Eastern cultures somehow saw the world as having emerged from water, no-one before the Greeks had considered *substance* as a concept embodying both physical and spiritual potentials.

The first philosopher to do so – and who is generally regarded as the first speculative thinker in Greece – was Thales (*c.*625–547 BC), of the Ionian island of Miletus. Thales argued that water must, indeed, be the primary root of all things, for it could appear as a liquid, a vapour, or a solid. The Sun could clearly suck it up into the sky for it to descend as rain, in addition to which it was essential for all life. Water, in fact, was the first clear expression of an 'element', preceding Heracleitus' fire by several decades.

Then Anaximenes (*fl.*546 BC) – the last of the Milesian philosophers – proposed that air was the fundamental stuff from which all things were made, for air supposedly expanded and contracted, and somehow changed its physical form. Air, Anaximenes asserted, must be the root of all things, because without air there could be no 'breath of life'; and one glimpses here an interesting parallel to the *Genesis* creation story, where all cognisant life is suffused with God's breath.

However, Thales' own pupil, Anaximander (*c.*611–547 BC), had doubted that any one single substance could lie at the heart of everything that exists, and had argued instead for a primary, indefinite property which he styled *apeiron* ('boundless', 'infinite'). This concept of a primary, adaptable stuff from which all things could be formed was later to be developed by Leucippus and Democritus, some time before 400 BC, as the first atomic theory.

The most all-embracing and historically enduring of ancient theories of substance, however, were the four elements of Empedocles (*c.*490–*c.*430 BC). Thales' water, Anaximenes' air and Heracleitus' fire were complemented by earth to create the versatile properties of wetness, dryness, hotness and coldness, the constant interplay between which, argued Empedocles, could give rise to every substance and state of being. When in balance, these elements produced rest, but when one usurped the place of another, the elements raged together to produce chaos. They even lay at the centre of the interpretations of health and disease and the explanation of human temperaments in subsequent Greek medicine, as well as being fundamental to the philosophies of Plato and Aristotle.

But instead of pursuing the idea that a single substance or set of four properties was primary to everything, other philosophers sought for a foundational truth that lay in numbers and geometry, which was to lead to another of Greece's great contributions to the history of human thought. Pythagoras of Samos (*c.*580–500 BC) was clearly impressed by Thales' concept that nature was a unity, and was the first to clearly articulate the notion that *number* lay at the heart of all things. According to legend, he did this after hearing the different pitches of sound produced by large and small hammers on a blacksmith's anvil. Sounds striking the ear could be equated with given metallic weights, in the same way that given lengths of vibrating string or blown pipe always produced the same pitch. What is more, the weights or lengths of string or pipe required to give a particular note always possessed the same arithmetic ratios to each other, thereby confirming that, amidst the constant flux of the natural world, there were eternal and unchanging properties. These musical sequences also related to an aesthetic that lay implanted in the human intellect, for certain proportions of sound were pleasing and others made people grind their teeth. It seemed, therefore, that arithmetic, weights, linear proportions and the rational intellect were somehow all bound together in one beautiful unity.

Not only did Pythagoras consider terrestrial objects as so related, but also the planets. Indeed, he conceived of a unified cosmos, in which the proportionate velocities of the planets, and their presumed distances from the 'central fire' around which they were believed to turn, constituted exact and eternal mathematical harmonies.

To Pythagoras and his followers, numbers took on a sacred status, for amidst all of the flux of the world, they were the things that never changed. These numbers, moreover, did not relate to any particular object. Instead, they represented abstract pieces of eternal perfection that could be applied to any concept or solid object, be it a circle, a triangle, bags of olives, coins, or sounds. Sadly, no written works of Pythagoras are known to have survived, but legend has it that he also explored the geometrical proportions of circles and triangles, while the 47th proposition of Euclid's *Elements* (*c*.280 BC), which provides proofs for the proportions of a right-angled triangle, has been immortalised as the 'Theorem of Pythagoras'. Pythagoras and his followers, furthermore, were the first to recognise that the Earth is indeed a sphere hanging in space, and is not supported by anything.

It is clear that by the time of Plato, the Greeks had already developed a series of concepts that would be pivotal in the subsequent history of human thought. One can see, therefore, that the creation narrative related in Plato's dialogue *Timaeus* is fundamentally different from all creation myths that preceded it, although it is clearly influenced by the mathematical philosophies of the Pythagoreans.

Like the Pythagorean cosmos, Plato's *Timaeus* is about balance, number, and harmony, as one might expect from a dialogue named after one of Pythagoras' own disciples. It explores the concept of the 'music of the spheres', according to which the planets rotate around a now central Earth in precise geometrical ratios based on the cube, tetrahedron, icosohedron and octrohedron to create an harmonious whole: a concept, indeed, which was to provide a major incentive to the researches of Johannes Kepler 2,000 years later.

One of the most enduring themes in *Timaeus* is the way in which the Creator, or demiurge, does not make everything from nothing, but gives *form* to chaos. Things exist because the Creator has embossed order upon them, in the way that a potter makes a pot, or a seal impresses a design upon clay. But as the clay, or chaos, already exists and has characteristics of its own, the physical creation can never be perfect. Even so, there must still have been a philosophical necessity for the cosmos to come into being, and to bear marks of divine reason and number. As a necessary component of this creation is mankind, the human mind and body must have harmonious resonances both with the perfect realm of eternal ideas and with the ever-changing realm of physical substance – for just as the Creator has arranged the Empedoclean elements of earth, water, air and fire into ordered sequences, so the healthy human body has its equivalent four humours of black bile, phlegm, blood and yellow bile.

Timaeus, in conjunction with other Platonic dialogues, presents a coherent and beautifully argued statement about man, nature, and the epistemology of ultimate truth. It draws on Parmenidean ideas of truth as an immaterial and unchanging thing that is apprehended only by the intellect and not through the senses. It was also Pythagorean insofar as it saw number and mathematics as embodying these truths, for the Creator had built a world that was geometrical at its core. And in the character of Socrates (*c.*469–399 BC), who had been Plato's teacher and inspiration, one encounters the quintessential Greek philosopher.

Socrates was versatile, ingenious, a brilliant dialectician, a professional academic, and an 'idealist' who believed that truth was to be found in the eternal 'forms' that lay beyond everyday reality. As a young man, Socrates had displayed valour as a hoplite soldier in Athens' wars, and at Potidaea in *c.*431 BC (a battle fought against the Corinthians, another Greek city state), Socrates saved the life of Alcibiades. Even on campaign, however, his mind was on philosophy in the rest periods between battles. Indeed, it was in this military stage of his life that, according to a story related in Plato's *Symposium*, Socrates' extraordinary powers of intellectual abstraction became apparent; for on one occasion he became so engrossed in a problem that he stood stationary for a whole day, so that his body acted as a sundial. Only when he had found a solution did he move! According to Plato and Xenophon, he understood geometry, and it was most likely due to the seemingly absolute truths implicit within it that Socrates followed a Parmenidean path that led him to try to pass through the 'illusions' of everyday experience to the eternal forms or ideas beyond.

Central to this method was geometry. Indeed, it is impossible to overestimate the importance of geometry to the post-sixth-century BC Greek intellectual world, for it was to become the archetype of all demonstrable truth, and therefore above argument or opinion. It is difficult to determine exactly how this geometrical culture arose in Greece. It is certain that the Greeks were not the first people to lay down lines and triangles, or to divide circles, for the Egyptians were expert surveyors and architects, while the Babylonian zodiac, amongst other things, indicated a practical familiarity with the geometrical properties of circles. The prior existence of Babylonian place numbers and computational techniques, and a possible knowledge of spare roots, moreover, also clearly show that the Greeks did not invent mathematics.

Yet what the Greeks did with mathematics was profoundly radical: they gave it a new philosophical power. Beginning with a collection of useful computational techniques – probably derived from the Babylonians and the Egyptians – the Greeks began to explore a conceptual world of numbers, the properties of which were eternal and unchanging. This was the first successful penetration of the veil of shadow, illusion, and chaos, and promised to lead to the realm of perfection and absolutes.

Much of the very early Greek geometry – especially that which was reputed to date from Thales and Pythagoras in the sixth century BC – has a legendary quality about it, and as none of the early geometers left written treatises behind them, we know of their exploits only from the pens of their successors. One such legend about the Pythagoreans (written down, it is true, many centuries later) says that when they obtained a definitive proof of a particularly knotty theorem, they sacrificed oxen as a gesture of thanks to the gods. Yet it is clear that not only had geometry achieved a definitive, impeccably reasoned and intellectually unchallengeable status by the time that Euclid composed his famous *Elements of Geometry* around 300 BC, but that its theorems and premises had already been central to the thought of Socrates and Plato more than a century earlier. Indeed, it was even said that above the doorway into the celebrated Athenian Academy, which Plato himself was to head, was engraved the inscription: 'Let no man ignorant of geometry enter.'

The theorems of geometry constituted the perfect examples of those ideas that were to be immortalised by Socrates and Plato in the doctrine of *forms*. The real truth, of course, did not subsist in the triangles and circles that one drew with instruments upon a board, for by definition of having been drawn and made physical they were imperfect. At best, human hands and tools could only approximate to the truth. What really mattered were the *forms* or *ideas* of the triangles and circles, which lay beyond the drawings. For neither by demonstration nor by logic could one conceive of a triangle the internal angles of which did not contain exactly 180 degrees, or half the degrees in a circle. Nor could a straight line that ran through the centre of a circle do anything other than bisect the circle. These forms or archetypes, moreover, did not exist in any particular place, but resided in the eternal 'mind' of the Creator, as potential waiting to become actual.

Logic sprang from the same source, for when one used the correct definitions, and constructed syllogisms based upon a dialectical process of argument and counter-argument, it became possible for the human mind to find truth as an absolute. It was through this process of analysis and precise definition, therefore, that the Greeks aspired to break the ever-spinning wheel of chaos to enable the philosophical intellect to know truth and beauty in all their glory.

This way of thinking permeates all of the Platonic dialogues. In the *Republic*, Socrates explores the necessary conditions of an ideal *polis* or state, based as it must be upon every man occupying his right place; and in the *Meno*, he talks about how we can aspire to true philosophical understanding, by showing that geometrical knowledge lies latent in the minds of all men, if only we choose to bring the latency to actuality. Indeed, in the *Meno*, Socrates addresses questions about the proportions of triangles to a young uneducated slave-boy, and by a process which might look suspiciously like leading the witness, extracts several important concepts from him. It was Socrates' opinion that in a former life (he believed in reincarnation) the soul of the slave-boy had been conscious of the truths of geometry, and that when it came to being made aware of absolute truths, education was really a process of recalling or remembering that which was already embedded in men's individual psyches.

Likewise, the concept of Platonic love sprang from the same source. While goddesses like the Egyptian love and fertility goddess Hathor and her Greek counterpart Aphrodite were associated, amongst other things, with *eros* or sexual desire, Socrates argued within the realm of eternal forms and ideas for a higher and more perfect love which was not concerned with human bodies so much as with an aspiration to perfect truth and beauty. It is perhaps ironic that in Plato's and Socrates' world of cultured gentlemen – where women were generally *persona non grata*, and this higher contemplative love was usually shared between men – it was, according to Plato's *Symposium*, the prophetess Diotima who first made Socrates aware of it. In our modern-day society, in which physical sexuality has been elevated to the level of a monomania, and in which the spiritual is marginalised, it is important to put Greek Platonic love into perspective. This was a world in which powerful intellectual friendships between gentlemen were considered to be natural and proper, and in which the particular emphasis lay upon the intellectual, not the physical. And while, in the nature of things, some of these friendships no doubt contained marks of physical affection, *eros* was not their intended end. To see Platonic love as somehow prefiguring our modern concern with gender orientation is to distort the past.

In the *Phaedo* – the last of the dialogues in which Socrates is the star – we not only have one of the most riveting death scenes in the whole of world literature, but the apotheosis of the philosophical life, an analysis of civic responsibility, and a powerful argument in favour of the immortality of the rational human soul, with a willingness to submit to death rather than acquiesce in tyranny and falsehood. Socrates the man was condemned to death, ironically enough, by a popular court, for supposedly teaching spiritual liberty to his pupils and friends. Deciding, however, that having the blood of the famous Socrates on its hands would hardly do the state credit, it laid plans to facilitate his escape. However, such an escape would have been cowardly, and so Socrates insisted on receiving his due sentence, much to the presumed embarrassment of the authorities. He casually sipped the cup of hemlock (the Athenian state poison) while talking about truth and immortality with his grieving friends. Then his feet started to go numb, but he still kept talking, and only when the drug allegedly reached his heart did he fall into his final sleep.

And yet Greek, and especially Athenian, culture, with its great concern for civic and intellectual freedom, meant that playwrights and wits could say things which provide us with alternative insights into the lives of its prominent figures. The image of Socrates which comes down to us from Plato and Xenophon, for instance, is deeply reverential. On the other hand, Socrates is sent up outrageously in

Aristophanes' play *The Clouds* (423 BC), in which the all-knowing philosopher is portrayed as something of a windbag. We also know from other sources that Socrates had a wife named Xanthippe, who was the daughter of a well-connected family, and that they had three children. We do not know, however, whether Xanthippe, in real life, was such a shrew as legend has made her.

In addition to the idealised Platonic love between gentlemen that Socrates praised, the uproarious Aristophanes reminds us that active homosexual relationships existed. In *The Clouds*, he hints that these are rife across the social spectrum, applying the term 'wide-arsed' (*euruproktos*) or 'loose-arsed' (*lakkoproktos*) to poets, politicians, lawyers, and even those sitting in the audience.

Trade, political independence, theatre, literature, and an abundance of leisured professional intellectuals made the Greeks a unique people in antiquity. And not only did these circumstances provide them with the mental space and inclination to wrestle with philosophical absolutes such as creation, geometry, and the noble life, but also to attempt to fathom out the very working structures of the starry heavens.

In addition to any encounter which the Greeks may have had with Babylonian mathematics, their very seafaring character brought certain physical truths home to them. In addition to their awareness of the geometry of eclipses (mentioned above), they quickly came to realise that from Egypt one could see, low down in the southern sky, stars which were not visible from Macedonia in northern Greece. Not only was the Earth spherical; the sky was also spherical. Quite simply, the more of the Earth one traversed, the more one became aware of the relationship between geography and astronomy.

As far as can be ascertained, the Greeks were the first people to study the movements of the heavenly bodies not for purposes of divination, but for sheer intellectual curiosity. Astrology was certainly practised by the Greeks, and Claudius Ptolemy's *Tetrabiblos* (*c*.AD 150) was to serve as its definitive compilation down to the sixteenth century AD, but astrology was not the driving force behind the science in Greece.

If there was one overwhelming concern running through Greek astronomy it was an attempt to understand planetary motion. Why do the planets move across the sky? Why do they do so in such complex orbits that in some cases decades have to pass before they return to the positions at which they were originally observed? Even as far back as Thales and Pythagoras, the Greeks were seeing those planets not so much as 'gods in the sky' but rather as natural objects that moved in a perfect, unchanging realm, and with orbits which should be capable of precise mathematical expression.

Central to the understanding of planetary motion was the mechanism that caused solar and lunar eclipses, and although (according to Herodotus) Thales had successfully predicted the solar eclipse of 585 BC, it is difficult to know how this would have been possible at such an early date. A scientific, as opposed to a speculative, explanation of planetary motion depends upon the existence of long runs of accurate observational data of the Sun, Moon and planets as seen against the fixed stars, thereby making it feasible to detect the presence of long-term cyclical patterns – and the only people in the sixth century BC who had such data were the Babylonians. Thales might have had access to Babylonian data, but the first true breakthrough in Greek 'celestial mechanics', or planetary motion studies, came with Meton around 430 BC.

Meton probably had access to Babylonian observational data, and some acquaintance with Babylonian knowledge about the cycle in which 235 lunar eclipses fit fairly closely into 19 solar years, or 6,939 days. This 'Metonic Cycle' provided one of the first rational mechanisms whereby calendars could be kept in synchronicity with the heavens, using a more refined technique for intercalating corrective days. It was subsequently employed by Jews for calculating the date of Passover, and by Christians for determining the date of

Easter, as well as providing useful criteria for the better calculation of eclipses.

Greek celestial mechanics, however, always operated within certain philosophical presuppositions that derived from their concepts of perfection and eternity; for while the terrestrial realm might be a place of transient illusions, the heavens embodied the true eternal Forms. This imposed two constraints upon astronomy that were explored by Plato in his *Timaeus*, and were still being wrestled with by Johannes Kepler in 1610. The first of these was that the heavenly bodies must move in perfect circles because, in geometry, the circle was a perfect shape. Secondly, the planets must move in their perfectly circular orbits at perfectly uniform and unchanging speeds. Perfect circles and uniform velocities, therefore, were intellectual prerequisites for all Greek astronomers.

But in reality the planets do *not* behave in this way. The Moon's orbital velocity around the Earth, for instance, was known to vary slightly in accordance with a cycle, while some but not all total solar eclipses failed to obscure the Sun completely but left a thin annulus of sunlight still visible. Surely this could occur only if the Moon's distance from the Earth varied slightly. And most puzzling of all were the retrograde motions of the planets – especially Mars, Jupiter and Saturn – which could be seen moving along in a uniform direction against the background of the fixed stars, but would then would stop, loop backwards, stop again, and then continue forward.

The Greek astronomers were by no means the first in antiquity to be aware of this problem, but they were certainly the first to devise a coherent explanation for it, which they did by devising and testing 'models' for the way in which the Universe might work. In this respect, the Greeks created yet another precedent in the history of human thought, for scientists today constantly 'model' nature by trying carefully considered hypotheses and testing their predictive potential against the best available observational data. Indeed, this modelling and testing is now used in every branch of science, from models of the cosmological Big Bang to models for the genetic coding of biological data.

The model which the Greeks devised for explaining the planetary retrogrades within the requirements of uniform speed and circularity was to imagine the cosmos as consisting of a series of perfect transparent crystalline spheres, each nestling within the others, rather like the skins of an onion. This model was first formulated by Eudoxus of Cnidus – a contemporary of Plato – in the fourth century BC. Eudoxus postulated a transparent sphere rotating west to east (the normal direction of the planets as seen against the fixed stars) on a polar axis. To the equator of this sphere a planet is attached. This planet-carrying sphere, however, has polar points that are anchored into a larger crystalline sphere, which completely encloses the inner sphere. But this outer sphere rotates east to west, and, what is more, its own pivotal points are set at an angle to those of the inner sphere. Consequently, the two spheres rotate eccentrically to each other, and when this motion is combined with their contrariwise direction of rotation, the path traced by the planet, as seen by a person set at the centre of both spheres, is a figure 8. This figure-of-eight shape was called a 'hippopede', or 'horse-fetter', as it resembled a device of equestrian control.

Some of the planets, however, exhibit more complex retrogrades or more complex orbits than do others, and Eudoxus therefore suggested that more than two internally pivoted spheres were necessary to account for their motions. The Moon, for instance, required three concentric spheres. The outer sphere controlled the Moon's east-to-west orbit around the Earth, whereas the other two were used to account respectively for the lunar month and the 18.6-year cycle that was connected to the occurrence of eclipses.

But did Eudoxus believe that the sky actually contained these multiple spheres of crystal nestling inside each other? Or were they perhaps an intellectual device invented for explaining a complex piece of observed natural phenomena in accordance with the necessary philosophical principles of perfect circularity and uniform speed of motion? We do not know – but that is not what matters. What *does* matter is the appearance, for the first time in intellectual history, of a new way of looking at nature. In the fourth century BC, Eudoxus and his contemporaries, in their use of hypotheses and mathematical models, were venturing beyond empirical calculating techniques on the one hand, and fanciful speculation on the other, to produce a rational basis for astronomical prediction. And even if subsequent generations of astronomers such as Aristarchus, Ptolemy or, 2000 years later, Nicholas Copernicus, replaced Eudoxus' 'hippopedes' with epicycles, eccentrics and other devices, the same quest for a comprehensive understanding of natural phenomena within the guidelines of exact mathematical description still remained unchanged. And it still does so today.

That wind which first blew across Attica around 600 BC was destined to bring about some of the profoundest changes in human culture. Upon a foundation of political independence based primarily upon commerce, the Greeks were to initiate an intellectual revolution that reinterpreted the divine not so much in terms of a pantheon of peevish deities but as an eternal and enduring universal intelligence whose very rationality was both the archetype and driving force behind all things. And along with that new understanding, philosophy was born, raising mankind from a subject race to become an active participant in the business of intellectual, artistic and social creativity. For now human beings felt as though they too could explore concepts such as truth, beauty, eternity, purpose and order, which the earlier Near-Eastern cultures had believed to be the preserve of their gods and goddesses.

From this movement, science in all its forms was also born, as the ordering force of the *logos* and the *nous* was glimpsed by human minds, and reason found a way of transcending chaos by logic, argument and mathematics. If Socrates and Plato embodied the Parmenidean approach to truth in their doctrine of eternal 'Forms', so Plato's pupil Aristotle drew inspiration from Heracleitus' ideas of change and becoming in nature, to emphasise the importance of physical phenomena in their own right as a valid domain of enquiry.

FIVE

Aristotle and his successors

One of the great turning points in the history of ideas came with Aristotle (384–322 BC), for while his predecessors – from Thales onwards – had pioneered radical new thinking in mathematics, geometry, and the understanding of the human condition, the majority of these philosophers worked within an idealist tradition, in which the mind, and its unique ability to apprehend the perfect Forms that lay behind the illusion of the physical world, was all that really mattered. But Aristotle modified this somewhat. While he agreed with his Athenian mentor, Plato, that Forms might be the ultimate and eternal truths, he nonetheless believed that the physical world contained consistencies which were worthy of study in themselves.

Aristotle's personal background probably coloured his views in this direction, for he was the son of a medical doctor from Stagira in northern Greece, and had himself probably received some medical training. Indeed, a biological or interactive systems approach to problems, rather than an exclusive concern for a singular eternal truth, penetrates the whole of his thought; for in his investigations of natural phenomena he was less occupied with geometry and Forms than he was with observable patterns and structures in the physical world.

His voluminous writings – either penned directly by his own hand, or else recorded, it has been thought, by his students from his lectures and seminars – testify to the extraordinary breadth of subjects into which he conducted his researches. These subjects included astronomy, meteorology, optics, a wide range of biological and medical topics, psychology, and logic. It was also from the writings of Aristotle that such words as biology (from the Greek *bios* – 'life'), meteorology (from *meteoros* – 'in the air'), psychology (from *psyche* – 'conscious self', 'soul') and physics (from *physis* – 'nature as an originating power') entered international scientific usage, in much the same way that from his clinical predecessor, Hippocrates of Cos (*fl.*400 BC), scientific medicine developed a vocabulary based on Greek words for specific body parts and their diseases – a vocabulary which is still used today.

Like all the Greek thinkers, Aristotle was concerned with the nature of change and the tendency of chaos to upset order; yet unlike almost every other philosopher, including Plato, he did not seek a solution through a form of creation mythology. In fact, Aristotle argued that human beings can never possess certain knowledge about the beginnings or endings of creation, for as Heraclitus had pointed out, human experience reveals only constant change. The world, indeed, is in a continual state of motion, or becoming, which indicates a cyclical character to events that must preclude any certain knowledge as to what came first or who made it. Nor do the processes of the observable world tell us how long the creation will last. And as a pragmatic thinker who, like most doctors, is concerned with tangible truths, Aristotle devoted his energies to what lay before him. It is therefore easy to understand why some scholars have thought of him as the first 'scientist', as opposed to philosopher. Some of these traits in Aristotle's writings required some ingenious reinterpretation in subsequent centuries, to make them acceptable to Jews, Christians and Muslims who, from their religious traditions, *did* have a clear-cut idea of the origin of things.

Aristotle dealt with the problem of change through the explanation of two concepts: a development of the Empedoclean doctrine of the four elements, and his own ideas of potentiality, actuality and causality. Empedocles, it will be recalled, had argued that there were four elements that made up the world: earth, water, air and fire. In his *De Caelo* (*On the Heavens*), Aristotle identified these elements not only with four substances, but also with eternal properties: earth was heavy and cold, water was fluid and moist, air was elastic and dry, and fire was volatile and hot. And when they were mixed together, they could produce every complex structure under the Sun – from human flesh to wood, depending on the proportions of the intermixing. In fact, this intermixing could be observed directly when something was broken down, especially if it were set alight.

When a piece of wood burns, for instance, it will first spit, crackle and steam, as the *water* is liberated from it. Then will come the smoke or *air*, followed by the tongues of *flame* that leap up from the wood, seemingly from nowhere. And when all the burning is over, a grey inert ash will remain. This is *earth*.

And was it not possible to discover the natural abodes of these elements once their forced union in a piece of wood or other matter had broken down? Did not air always billow around one, whereas fire always rose straight up? And did not earth always sink to the lowest possible level, whereas water rested upon earth and yet lay below air? In his doctrine of the elements, therefore, Aristotle was able to assign specific places in nature to physical properties. These elements, moreover, had natural affinities and enmities amongst themselves. Water and fire, for example, were not compatible, and when air was trapped in earth, the ground rumbled and shook, producing earthquakes and volcanic eruptions.

It became feasible, therefore, to see in these elements a set of eternal principles which could conjoin and dissolve to produce all the ephemeral things of the world – growth and decay, the seasons, birth, death, and regeneration. These elements were in a constant state of *becoming*, and thereby formed the basis of Aristotle's idea of potential and actual.

For if all things are in a state of becoming, then any one thing must be the potential from which a future actual thing will spring. An acorn, for instance, is potentially an oak tree. The tree is potentially a piece of wood from which a chair may be fashioned, and the enjoyment of the chair is the end in itself, as far as earthly joys can go. This process of purpose and end, in fact, can itself be broken into four causal sequences. The *material* cause is the wood from the tree; the *formal* cause is the design of the chair that pre-existed in the carpenter's mind; the *efficient* cause is his imposition of order on his raw materials; and the *final* cause is the wish to sit in the chair.

But a final cause in one context can become the first or material cause in another, for if one were freezing to death, one would desire a fire. One might then choose to chop up the chair for firewood, in which case the desire for heat itself becomes the new final cause. And will not the ash form a soil nutrient to turn present acorns into future oak trees?

What Aristotle had achieved in the above was the taming of chaos without the need for a personified creation myth. Change was rendered sequential and amenable to rational explanation. Order prevailed when the four elements occupied their proper places, and chaos broke loose when they usurped one another's places. It was not for nothing, therefore, that in physics, medicine or politics, balance was the condition to which to aspire. This was to give rise to Aristotle's celebrated Doctrine of the Mean: the 'middle way' between extremes and excesses was the natural one to pursue if good order was to prevail.

Furthermore, it was the goal of medicine to keep the body's four humours of black bile, phlegm, blood, and yellow bile (the human equivalents of earth, air, water and fire) in balance, and both diagnosis and therapy hinged upon identifying and correcting the misplaced humour. And in politics, it was the duty of the wise lawgiver to maintain harmony by ensuring that each man occupied his correct social place, depending on his wisdom, moral virtue, and ability.

This approach runs through all of Aristotle's writings, and one source of his enduring appeal down the centuries – not only to later Greeks, but also to Jews, Christians and Muslims – was the marvellously integrated character of his thought. For while one might argue that Plato's thought, based as it was on the universality of perfect Forms, also possessed a power of integration, much of Aristotle's attraction stemmed from the fact that his philosophy did not dismiss the real world as a passing illusion, but treated it seriously and grappled with its changes. It contained a dimension of pragmatic reality which could be used (within its inevitable observational and experimental limitations) to explain all kinds of phenomena. Comets, for example, occurred in the upper atmosphere, where foul vapours were ignited by the realm of fire that surrounded the Earth. Stones fell when dropped, because they wished to return from the airy element in which they were suspended to be with their own kind in the ground. Yes, the elements had their own natural 'desires': not conscious desires like those of humans, nor instinctual desires like those of beasts, but an innate tendency to be with their own sort. In short, they possessed their own physical specificity. And what the elements most wanted was rest, which was possible only when they were with their own, although that teeming turbulence which characterised all terrestrial things forced them into inevitable conflict, change, and becoming; because in the world, motion was the ultimate reality.

In Aristotle's thought, physical perfection existed only in the realm of the astronomical bodies. In this respect, he was at one with his predecessors. But it was in his explanation for the same that he was to be significant, for to Aristotle terrestrial and astronomical bodies were made of different kinds of matter. On Earth there were the warring four elements; but the Sun, Moon, planets and stars were all made of one stuff: the fifth element, quintessence. And because of this unity of substance, there was no ground for conflict, and so consequently the heavens displayed eternal perfection.

Central to Aristotle's cosmology (at least, insofar as the terrestrial realm was concerned) was physics, or his treatment of the causes of motion. This ceaseless process of cause and effect was set going, at some indeterminately remote time in the past, by an 'unmoved mover' – an agent whose other characteristics we can never know by reason. Earthly motion was essentially the product of conflict, as things desiring rest responded to those pushes and shoves that were inevitable parts of the process of becoming. Earthly motions, moreover, came in two forms: natural and violent. A natural motion took place when an element returned to its natural place, such as when a stone fell to the ground. A violent motion existed when an object went away from its own kind. An arrow left a bow, for instance, under the force of a violent motion, but as this force weakened, the wooden ('earthly') arrow assumed a natural course, and fell to the ground.

Such violent motions did not exist in space, for in this realm of the fifth element, only natural motions were followed, and this meant that – in accordance with prevailing canons of astronomical perfection – bodies moved at uniform speeds in perfectly circular orbits. They also rotated around the Earth. Aristotle's science, indeed, was to anchor the geocentric cosmology into the Western scientific mind for the ensuing 1,900 years. He did so, and it stayed there, because geocentricism made eminently good sense in the light of available scientific knowledge. For do not objects fall to Earth because the Earth occupies the centre of creation? Does the Earth wobble, shake, or show any other signs of motion? And is it not eminently logical to think of the only changeable thing in creation as equidistantly placed from perfection, by being in the middle of it?

To Aristotle, the celestial and terrestrial realms were by definition polarised and not in communication with each other, for here there is no concept of gravity or any kind of all-pervading physical force permeating the entire Universe. Indeed, to Aristotle 'gravity' was synonymous with the heaviness of terrestrial bodies, and had no astronomical connotations. Gravity belonged to the realm of *physis*, or 'natural things' and their motion. And while that Greek word gives us our modern word 'physics', we must not forget that it was only after AD 1600 that Galileo and Isaac Newton realised that the same physical force – later identified not with weight but with *mass* – really does pervade the entire Universe.

While Aristotle's cosmos was sharply divided between the terrestrial and celestial realms, there was a natural agent that bridged them: light. This was not the light generated by terrestrial flames, but the pure white light of the Sun and stars. This light came down from the celestial regions, and after crossing the 'sphere of the Moon' or of the Moon's orbit around the Earth, entered the terrestrial realm of change and decay.

Because it was celestial, it was naturally white – the ancient colour of purity – and was also believed to be instantaneous in transmission from its source to the observer's eye. Yet once light encountered terrestrial corruption, it 'decayed' into colours. Aristotle's *Meteorologica* describes how this decay results in red sunsets, rainbows, and other aerial phenomena. When the light is passed through water, crystal or glass, it produces the familiar spectrum – red at one end and blue at the other. Aristotle's decay theory of coloured light would be accepted without question until (as mentioned in Chapter 1) Isaac Newton's brilliant series of experiments in the 1660s completely turned it about by showing the colours to be eternal and natural, and white light to be a mixture of colours. Aristotle also devised a theory of ocular vision, in which this light passed into the eye to form an image, and while his theory was wrong by modern standards, it provided us with one of the earliest models of ocular perception based upon the anatomy of dissected eyes.

Aristotle's writings sometimes strike us as containing an odd mixture of acute observation, insight, and sheer speculation. One of the reasons for this lies in the fact that Aristotle was not a scientist who perceived the importance of instruments when it came to studying the natural world. His observations concerned the construction and operation of things, rather than their precise measurements, for in his time the medical, meteorological and compositional aspects of nature that most interested him were not susceptible to measurement. After all, concepts such as blood pressure and the laws of chemical combination still lay millennia in the future, and an understanding of nature based upon the carefully observed innate tendencies of things seemed much more logical.

Aristotle's approach to science, therefore, was based on natural history observation and logical deduction rather than mathematics, and by 300 BC the Greek world was already generating scientific thinkers who would be attracted to one or other style of proceeding. And in no way were the two styles incompatible, for in matters of astronomy, let us say, Aristotle might explain why the single-element composition of the heavenly bodies made them perfect, whereas a Platonic mathematician would then go on to define that perfection in precise geometrical terms, and use it to predict eclipses. Indeed, it was the wonderfully integrated character of Greek science that made it so impressive and so influential. Down the ages, scientists in Alexandria in 50 BC, in Damascus in AD 1000, in Paris in 1300, and in Padua in 1600 would continue to claim to be either Aristotelians or Platonists.

While people used rules and balances for measuring everyday articles, the only properties of nature that could be accurately and

consistently measured in the ancient world were astronomical properties, and this was made possible through the Greek adoption and
extensive development of the Babylonian circle of 360 degrees, each
degree of which was subdivided into 60 minutes and then into 60
seconds of arc. And because the circle was a closed system, measurements made within it could be shown to be true or false by simply
adding them together to see if one obtained a total of 360 degrees.
Such measurements, backed by geometrical theory, gave science its
first insights into the idea of falsification: a system whereby one could
prove the truth or falsehood of a procedure by physical cross-checks.
This concept of cross-checking would later become one of the
cornerstones of scientific method.

Apart from the 360-degree circular motions of the heavens,
however, what other natural property could one reasonably measure?
A doctor might choose to ascertain his patient's height or weight, but
what could these reveal when it was the physically intangible humours
that governed health? Apart from the perfect heavens, one could
measure only the shifting sands of a teeming world of nature. It
required the development of a concept of systematic *experimentation*,
and a belief in the physical consistency of the natural world, before
such a radical application of measurement, and hence the extensive
use of instruments, could occur – and that would not happen until
after AD 1200.

The use of the 360-degree circle led to a series of remarkable
investigations in physical astronomy, as Greek scientists grasped the
interrelationship between precise measurement, geometrical theory,
and prediction. We have seen (in Chapter 4) how Eudoxus began to
'model' the heavens in an attempt to reconcile the observed retrograde motions of the planets with the philosophical criteria of
perfectly circular planetary orbits and uniform velocities. By the time
of the death of Aristotle in 322 BC, astronomers were becoming
extremely adventurous in their researches, and were even beginning

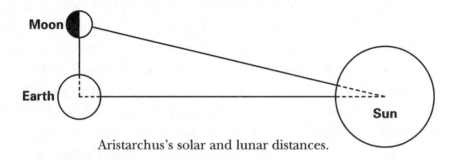

Aristarchus's solar and lunar distances.

to speak of giving dimensions to solar system bodies based upon angular observations and the use of geometrical theory.

Around 280 BC, Aristarchus of Samos made the first serious geometrical attempt to do this, as described in his treatise *On the Size and Distance of the Sun and Moon.* His method was based on the geometrical fact that when the Moon is exactly at quadrature (half Moon), then the Earth, Moon and Sun should form a long, thin, right-angled triangle, with the Earth–Moon line as the short side. But he discovered that this did not happen, for at quadrature the Moon makes slightly less than a right-angled triangle with the Earth and Sun. From this angular value he calculated that the Sun was nineteen times further away than the Moon. However, while the basic geometry of his method was correct, because of the difficulty of discerning, with the naked eye, the *exact* moment of half Moon, and due to his practical inability to measure very small angles with precision, his figure was erroneous. In fact, the Sun is about four hundred times more remote than the Moon. However, his attempt was bold and brilliant, and showed the way ahead for astronomical research.

Aristarchus's work hinged on the complementarity of angles within the triangle, and from the certain knowledge that if one knew the exact length of the base line of a right-angled triangle, and at least one other internal angle, then one could calculate the lengths of the other sides of the same triangle. The systematisation of this geometry was being undertaken in the thirteen-part *Elements of Geometry* by Aristarchus's exact contemporary, Euclid, of the Graeco-Egyptian city of Alexandria. It was not for nothing that Greek astronomical geometry was to develop so rapidly after about 300 BC. As Euclid dealt with 'plane' or the single-surface geometry of triangles, squares and circles, so Apollonius of Perga and later of Alexandria (262–190 BC) developed the three-dimensional geometry of the cone.

Cones are fascinating objects for geometers, for depending upon the angles at which a cone is cut through, it generates a series of curves. If it is cut at right angles to its vertical axis (the line between its point and the centre of its base), then a perfect circle is produced; but if the cone is cut by inserting flat planes through it at a variety of angles, then the result is an *ellipse*, a *parabola*, or a *hyperbola*, all of which curves have important geometrical properties. It was realised, for instance, that the projection lines of an ellipse closed to form a continuous curve that, unlike a circle, did not have a single centre, but

Apollonius's conic sections.

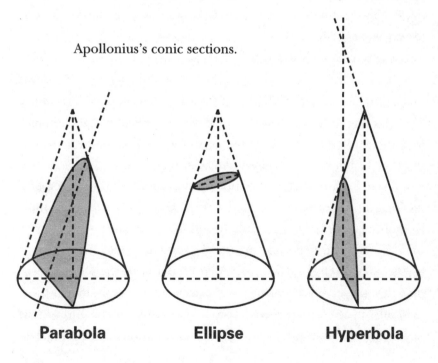

Parabola   **Ellipse**   **Hyperbola**

rather two focal points. On the other hand, both the hyperbola and the parabola were discovered to be 'open' curves whose lines diverge to infinity and never meet. It was also noted that these different curves had optical characteristics, and reflected and refracted light differently when imparted to mirrors and lenses; and in the seventeenth century AD, Johannes Kepler and Sir Isaac Newton eventually realised that planets and comets moved in orbits of these shapes.

Once again, that ever-present connectedness of Greek thought shines through, as the *logos* or *nous* brings into being and sustains an ordered and integrated creation, in which logic, mathematics, light, heavenly bodies, the four elements and bodily humours, motion, potential, actual, and becoming, form a whole to which the disciplined human intellect has access.

During the fourth and third centuries BC, in fact, several astronomers set themselves the task of trying to measure the physical dimensions of the Earth and heavenly bodies. Improving on a prior work by Eudoxus, the Alexandrian philosopher and scientist Eratosthenes (276–194 BC) made an ingenious attempt to measure the vital statistics of the Earth itself. Eratosthenes had heard that there was a deep well at Syene (modern Aswan) in upper Egypt, down which the noonday Sun shone, and was reflected in the water, at the time of the summer solstice in June. Yet on the same day, in Alexandria the noonday Sun cast a shadow which showed it to be 7 degrees 12 minutes of arc from the zenith.

From his knowledge of complementary angles and the geometry of circles, Eratosthenes realised firstly that this 7.2-degree angle comprised $\frac{1}{50}$ of a circle. He then employed a land measurer to pace out the distance between Syene and Alexandria, and found that it came to 5,000 Greek *stadia*, or stades. From this he was able to calculate that the spherical Earth's circumference must be 5,000 x 50, or 250,000 stades, and thence – from a knowledge of the proportions legendarily discovered by Pythagoras that the diameter of a circle will

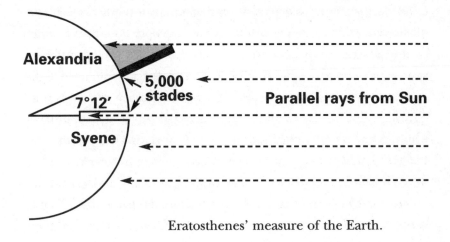

Eratosthenes' measure of the Earth.

divide 3½ times into its circumference – to extract the diameter and radius of the Earth.

Eratosthenes' method was impeccably correct, as far as its theoretical assumptions were concerned. However, we are not sure exactly which stade (of several in use in the ancient world) he used, nor do we know how much he 'rounded up' the measured distance from Syene to Alexandria to produce the suspiciously elegant 5,000 stades. Yet if he used the common stade which was equivalent to 606.75 British feet – as he could very well have done – then he would have calculated a terrestrial circumference of 28,728 British miles, which is by no means a bad approximation to the circumference of 24,902 miles that we accept today. If, on the other hand, we choose to use Eratosthenes' *revised* value for the Earth's circumference – 252,000 stades – and assume that he was using the *short* stade, then this translates into 24,662 English miles!

Of course, in addition to our uncertainty about his choice of value for the stade, Eratosthenes was ignorant of the fact that the Earth is not in reality a perfect sphere, but is slightly oblate, as its equatorial diameter is about 27 miles more than its polar diameter. It is, however, difficult to deny that his measurement of the Earth places him as one of the founders of the science of geophysics, and a pioneer of the idea that the theorems and concepts of geometry could be applied to the real world to gain new knowledge.

But it was two relatively late Greeks who were to give to astronomy its definitive classical formulation, and to supply it with a set of concepts and observational techniques which were to carry it through the next 1,600 years – right down, indeed, to the Renaissance. They were Hipparchus and Claudius Ptolemy.

Hipparchus's creative life, between about 150 and 125 BC, was spent on the Greek island of Rhodes, which was not only a great thoroughfare of the classical world, but also a manufacturing centre for ingenious artefacts of metal. Hipparchus's astronomical thought – like that of all the leading post-Eudoxan Greeks – was dominated by the problem of planetary motion. For what one sees taking place by 250 BC is a trend away from the philosophical quest to find the ultimate substance or agency of creation, which had occupied Thales, Pythagoras and the pre-Socratic philosophers, towards the development of a consistent mathematical model to explain how the heavens moved. In this development one suspects that two things had been of formative importance. One was the philosophical system of Aristotle himself, which had not only drawn out the best from what had gone on before, but had incorporated it into an original synthesis of knowledge that aspired to provide logically connected explanations for almost everything, from the whiteness of starlight to the reproduction of fishes and the causes of dreams. The other was the flowering and consolidation of geometry, with its immense power to demonstrate eternal truths that nothing could falsify, and to use them to analyse motion.

One of the ways in which Hipparchus applied geometry to astronomy was in his furtherance of Aristarchus's method for trying to measure the distances of the Sun and the Moon from the Earth.

For this, Hipparchus used the shadows cast during eclipses, when the Earth, the Moon and the Sun were in a straight line. Like earlier Greek astronomers, Hipparchus realised that one of the problems inherent in measuring the lunar distance from the Earth was the observed fact that the distance varied. This became evident, for instance, during total eclipses of the Sun, when the Moon passed between the Earth and the Sun (as we saw in Chapter 4). Sometimes the Moon blotted out the entire solar disk, causing a sudden temporary darkness to fall upon the Earth, while during others a thin annulus or ring of brilliant sunlight remained around the Sun, even at the moment of totality. This apparent variation must be caused by the Moon being very slightly more distant from the Earth at some times than it was at others.

From these shadows cast during eclipses, Hipparchus calculated that at its nearest to us, the Moon was only 59 Earth radii away, and at its furthest, it was 67⅓ Earth radii. In fact, his figures were astonishingly good, and enabled him to establish a reliable value for the

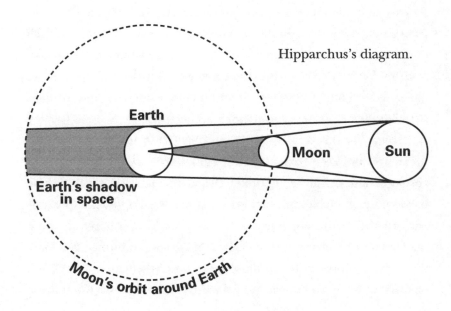

Hipparchus's diagram.

Moon's distance. Hipparchus also realised that if one could calculate the distance and the diameter of the Moon, and if during a total eclipse it could completely blot out the disk of the Sun, then there must be a clear proportionate relationship between the diameters and respective distances of the Sun and the Moon. Either the Sun was relatively small and not many times more distant from us than was the Moon, or else it was very large and an incredibly great distance away. In 150 BC there was no real way of knowing which was the case, but Hipparchus computed that it was at a distance of 490 Earth radii, which is only about one $\frac{1}{48}$ of the actual distance. Even so, by this date certain Greek scientists were becoming sufficiently confident in their methods to be willing to ascribe dimensions to the Universe; and they were doing so by using mathematical techniques which possess an intellectually valid connection to those which we use today.

Nowadays, however, we take it for granted that when a scientist sets out to measure the distance of something, he or she will do so in terms of clearly defined units such as miles or kilometres. One of the problems faced by astronomers of Hipparchus' time, however, was the absence of fixed natural standards of measurement, for while Eudoxus, Eratosthenes, and later Ptolemy, measured the Earth's diameter, each of them arrived at a different value because of the limited data available to them. Yet they all knew that the Earth was a sphere, and that spheres contained within themselves universal proportionate relationships between their diameters and their circumferences.

It was for this reason, therefore, that Greek astronomical measurements were always expressed in terms of proportions rather than in absolute values. Indeed, grasping the mathematical reality of these fixed and unalterable proportions in nature was an essential preliminary to the growth of science, for they wedded concepts in pure mathematics to observed physical realities, and confirmed not only the truth (as opposed to the illusory) status of the natural world, but linked that world to knowledge gained by the five senses when those

senses themselves had been properly disciplined by the rational intellect through a logical or mathematical training.

Indeed, a modern application of this method can be taken from
twentieth-century cosmology. In the early 1930s, the American
astronomer Edwin Hubble established, from spectroscopic observations and statistical analyses, a set of proportions from which he
claimed to be able to calculate the expansion rate and even the age of
the Universe. Over the intervening 70 years a vastly greater body of
observational data about stars and galaxies has been amassed than was
available to Hubble, as a result of which the actual vital statistics which
he ascribed to the Universe have been successively modified and
refined. Yet the basic formulation and proportionate relationships
within Hubble's Constant have stayed the same, just as did the proportions within Hipparchus's diagram of the Earth–Sun–Moon shadow
relationship, in spite of vastly more accurate data on the absolute
dimensions of these bodies becoming available in the meantime.

By the time of Hipparchus, moreover, it was becoming clear that
the heavens were not quite the same as they had been in previous
centuries. Even the stars themselves seemed to display great cycles, so
that what an astronomer saw in 150 BC was not quite the same as what
a predecessor of two or three millennia earlier would have observed.
This did not mean, of course, that the heavens were mutable or given
to corruption and chaos as were earthly things, but that there were yet
greater cycles for the astronomers to discover, using a combination of
exact observation, historical record, and mathematical analysis.

The use of these techniques led Hipparchus to make one of
astronomy's most fundamental discoveries: the precession of the
equinoxes. The two equinoxes (from the Latin – 'equal nights' (and
days)) occur in March and September, and are the two points at which
the Sun's annual passage through the twelve signs of the zodiac, along
the ecliptic, intersects with the imaginary circle called the celestial
equator, which is a projection of the Earth's equator onto the sky.

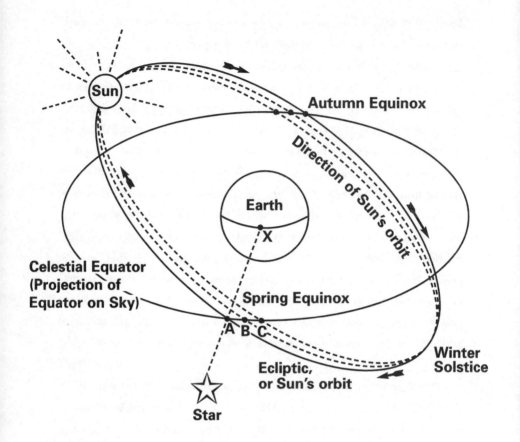

Sun

Autumn Equinox

Direction of Sun's orbit

Earth

X

Celestial Equator
(Projection of
Equator on Sky)

Spring Equinox

A B C

Ecliptic,
or Sun's orbit

Winter
Solstice

Star

Precession of the equinoxes. In any given
year, an astronomer at 'X' on the Earth's
surface will see the Sun at spring equinox
near to a star, along the line of sight
passing through 'A'. But the next year, the
Sun's path across the sky will have fallen
back – precessed – to fall behind the star to
'B'. And in the third year it will have fallen
back further, to appear at position 'C'; and
so on, through 25,800 years, when it will
have gone full circle to return again to 'A'.

These two circles – like a pair of rings drawn among the stars – intersect with each other at an angle of 23½ degrees, which is the angle of tilt of the Earth's axis (the obliquity of the ecliptic). This tilt, of course, produces the seasons, giving us (in the northern hemisphere) our summer solstice, or longest day, in June, and our winter solstice, or shortest day, in December.

What Hipparchus discovered, however, was that the points at which the zodiac and the celestial equator intersect – at the March and September equinoxes – do not stay in the same place with respect to the stars. From his analysis of earlier astronomical records, he found that the equinox points were sliding back (precessing) by a tiny fraction each year; and while his measured value for this annual precession was less exact than that available to modern astronomers, he nonetheless was the first to identify precession as a real phenomenon in nature. (We now know that it is produced by a very slow 'wobble' of the Earth upon its axis, and that it takes 25,800 years to precess through a full cycle.)

Yet why should Hipparchus's discovery be of such importance? Well, without it astronomers could not explain why celestial events described by astronomers three millennia earlier did not take place amongst the same stars as they do nowadays. (For example: in 3500 BC the helical (pre-dawn) rising of Sirius was a valuable predictor of the inundation of the Nile, but today it is no longer applicable.) Hipparchus defined his spring (vernal) equinox point as falling into the zodiacal constellation of Aries, thereby defining the beginning of the astronomical year from the 'First Point of Aries'. In the 2,150 years which have elapsed since Hipparchus's time, however, this equinox point has precessed out of Aries and through Pisces, and in the late twentieth century it entered the constellation of Aquarius, much to the inspiration of popular song writers, astrologers, and New Age dreamers who believe, for whatever reason, that 'the Age of Aquarius' will be replete with mystical significance.

Hipparchus's discovery of precession, however, was of scientific value because it also provided a yardstick against which the observations of earlier astronomers could be corrected, so that they could be compared with subsequent or even present-day observations. And from these observations it was possible to discover numerous small motions in the long-term Earth–Moon–Sun relationship without which Copernicus would have lacked the necessary data to develop his heliocentric theory in 1543, while Newton's knowledge of the solar and lunar orbits would otherwise have been too coarse to have enabled him to formulate the laws of gravitation.

An accurate knowledge of precession is also crucial if one wishes to compile a precise set of tables, or map the stars in their constellations, against which to record the paths of the Sun, Moon and planets as they move across the night sky. And Hipparchus not only made original astronomical observations of his own, but collected, corrected and compiled into accurate tables those of his Greek and Babylonian predecessors. From these he is said to have constructed, in 150 BC, the first astronomical globe and star maps, in which the equinoxes, solstices and other coordinates were related to specific stars, thereby rendering them of practical service not only to future astronomers but also to geographers, who would use the positions of the Sun and stars to map the locations of terrestrial topographical features.

The precession of the equinoxes is referred to as a 'constant', insofar as it never changes, although over the centuries, and especially after 1650, astronomers have refined the precessional value until we now know it with the minutest accuracy. And nowadays, in the age of digital computing, such knowledge is of use not only to astronomers, but also to historians, archaeologists and other scholars who deal with the past.

If, for example, a reference to an unspecifically dated historical event that is linked with an astronomical occurrence is being investigated, then an exact knowledge of precession is vital to programming the computer to 'back calculate' the heavens to the era of the event. The Star of Bethlehem, which St Matthew's Gospel in the Bible tells us heralded the birth of Jesus Christ, is now generally believed to relate to a brilliant conjunction (a close apparent mutual approach) of the planets Jupiter, Saturn and Mars in February, 6 BC. Indeed, in 1606 the German astronomer Johannes Kepler first suggested this possible explanation for the nativity star, for another such planetary conjunction had occurred in 1604. From his knowledge of precession and other astronomical constants, Kepler was able – through a mathematical analysis of earlier planetary observations – to establish that this particular type of planetary conjunction takes place every 805 years, with previous events occurring in AD 799 and 6 BC!

Indeed, a progressive understanding of the exact motions of the heavens, combined with a growing knowledge of constants or correction factors, has always inclined astronomers to try to date historical events from celestial criteria; and now that the computer has greatly eased the grind of calculation, and has vastly accelerated the process, this method is being used more and more.

For the Greeks themselves, however, the heart of the astronomical quest, with its constant accumulation of new data and its mathematical reconciliation with that from the past, lay in trying to explain how the planets moved around the Earth. From Eudoxus (whom we encountered in the previous chapter) onwards, this involved devising schemes whereby the transparent 'spheres of heaven' could correspond to observed reality within the intellectually assumed parameters of uniform orbital speeds within perfectly circular orbits. And it seemed that the solution, in part, could be found by establishing the correct number of spheres, some of which counter-rotated. In 350 BC, Eudoxus had postulated a Universe based upon twenty-six spheres, Callippus had added another nine, and Aristotle soon afterwards proposed between forty-seven and fifty-five. Then Apollonius, the conic section geometer, proposed an alternative to the multiplicity of spheres. Why not suppose, he argued, that while the Sun, Moon and planets rotate around the Earth in perfect circles, the Earth is not at the actual centre of these circles, but is offset, thereby making the circles eccentric? This slight displacement could be used to explain variations in planetary appearances as the distances varied.

Yet the most *avant garde* of all Greek planetary theories was that of Aristarchus of Samos, who around 280 BC proposed not an *Earth*-centred (geocentric) cosmos, but a *Sun*-centred (heliocentric) cosmos. For if one set the Earth in motion around the Sun, together with the rest of the planets, then one could explain the looping retrograde motion of the planets as line-of-sight effects, as some planets moved faster than others and appeared to overtake them. However, setting the Earth into motion seemed not only to contradict plain common sense, but also appeared to undermine the four-elements physics of Aristotle, which predicated that the Earth was the heavy and immobile centre of all things.

The reason why Aristarchus's heliocentric Universe did not seriously re-emerge until the time of Copernicus some 1,800 years later was not due to any deliberate suppression, but because ultimately, within the scheme of available knowledge in antiquity, it posed more questions than it answered – and that is never what one looks for in a scientific explanation.

Classical Greek astronomy reached its grand finale in the *Great Syntax* or *Great Compilation* of Claudius Ptolemy (*c.*AD 130–175), which came to be known to posterity under the title which the Arabs subsequently gave to it: *Almagest*. Ptolemy's *Almagest* brought six centuries of Greek astronomy together into one encyclopaedic digest. Not only did he use the recorded observations of his fellow Alexandrians and other Greeks to provide an historical database for his work, but he also drew heavily on available Babylonian observations extending back to the eighth century BC. In addition, he compiled the 'definitive' list of 48 astronomical constellations that would come down to us from the ancient world, and further provided decisive proof that the Earth was a sphere. Nor must we forget that Ptolemy also invented our modern practice of dividing the world into lines of latitude and longitude that correspond to precise angular positions among the stars, as he described in his other great treatise – *Geographia*.

Like all of his Greek predecessors, however, Ptolemy's overwhelming concern was to explain the motions of the planets – motions, indeed, which a growing body of accurate observational data had shown to be more complex than Eudoxus or Callippus could possibly have imagined. And like all mainstream Greek astronomers, Ptolemy worked within the context of an Earth-centred (geocentric) Universe.

His explanation for the various retrograde and variable distances of the planets lay in his development of a geometrical device used by Hipparchus and other Greek astronomers: the *epicycle*. An epicycle is a circle, the centre of which rotates around the circumference of a larger circle. Imagine a wheel rotating at a perfectly constant speed, and on the wheel's circumference a point. Around this point rotates another, smaller wheel, also at a uniform speed. Now attach, let us say, a small light to the circumference of this smaller wheel. When the large and small wheels are set into motion in the dark, this light will describe a series of perfectly uniform, regular loops. In short, we have replicated the backwards–forwards looping of the planets from a combination of two motions which, in themselves, are perfectly circular and uniform.

By the time of Ptolemy, however, the growing body of recorded astronomical observations available within the Greek world meant that this simple model of an epicycle – using as it did only *two* motions, both in the same direction – was inadequate to explain all phenomena. What Ptolemy did, therefore, was increase the number of epicycles, constructing circles with a succession of lesser circles rotating around points on their circumferences to finally account for the

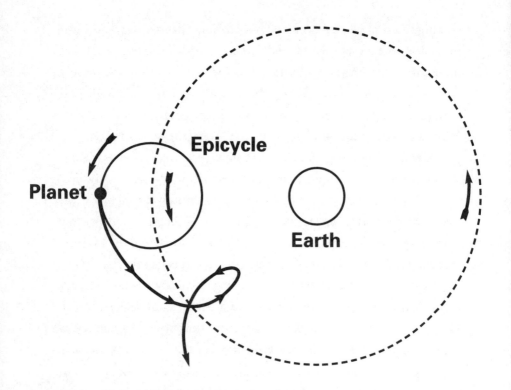

The epicycle. Two perfect
circular motions – the planet's
motion and the centre of the
epicycle – produce loops that
account for the observed
retrograde motion of the planet.

observed motions of each planet, rather like an increasingly complex
train of gear wheels. And of course, the last wheel, or epicycle, in each
train carried the actual planet that glowed in the sky. What Ptolemy
had done, in fact, was to model the Universe by inventing complex
mathematical configurations of epicycles to account for observed
planetary motions.

But how were all of the new data about planetary motion, upon
which Ptolemy drew to frame his theories, actually obtained? Here, in

fact, we encounter yet another major contribution of the Greeks to the history of scientific thought, for they were the first people in antiquity to realise the crucial importance of instruments when it came to understanding nature. For while the practised human eye, combined perhaps with the outstretched fingers of the hand, can give one a workable approximate yardstick with which to track the Sun, Moon and planets against the starry background, they can never afford precision. It seems to have been the Greeks, with their intellectual concern for numerical perfection, who developed the 360-degree circle from being a conceptual division of the zodiac and realised it as a mathematical instrument.

Certainly by the time of Euclid – and probably for several centuries before – Greek geometers had realised that a pair of compasses could be used to fit the radius of a circle exactly six times into its circumference, to form a hexagon. And as each of these one-sixth divisions always contained exactly 60 degrees, a careful compass bisection always produced 30 degrees. By adding 60 to 30, one could build up a perfect 90-degree right angle, quarter circle, or quadrant. Further bisections, trisections and quinquesections finally broke down the circle to 360 equal degrees.

Now if such a 360-degree ring was set up in the plane of the ecliptic (the Sun's annual path through the zodiac), then one could measure the positions of all astronomical bodies from the stars with great accuracy. And if the ecliptic ring was paired with a second 360-degree ring intersecting it at right angles across their mutual diameters, then one could measure vertical angles as well as horizontal angles. This configuration of rings was immortalised by Ptolemy as the armillary sphere, for when such a pair of rings, maybe six feet across, was fixed upon a rigid mount and aligned with the planes of the sky, and each ring was equipped with a pair of brass sights, then one could make very accurate measurements of the motions and retrograde motions of the planets. When these were reduced to a

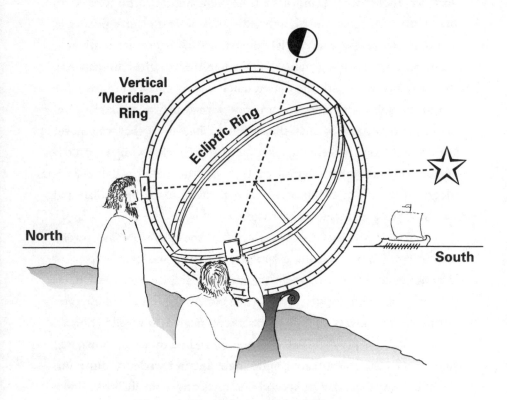

Vertical 'Meridian' Ring

Ecliptic Ring

North

South

The armillary sphere. Using sights that could be moved around the rings, astronomers could make accurate sightings of astronomical bodies across the centre point of the armillary. The astronomer on the left is measuring the vertical angle of a star above the horizon on the north–south 'meridian' ring. His colleague is measuring the east–west angle of the Moon on the 'ecliptic', or Sun's orbit ring. From their respective observations, the angle between the Moon and the star can be calculated.

common standard of measurement, and entered into tables or drawn up to form maps of the sky, then the astronomer had a rich treasury of hard, mathematical data on which to construct his theories.

What is more, the armillary sphere was not the only astronomical instrument devised by the Greeks. They used quadrants (quarter circles) divided into 90 degrees, with strings and plumb lines to measure vertical angles, such as that of the Pole Star; while Ptolemy devised a set of sliding rods – called Ptolemy's Rulers, or triquetum (three rods) – based on the complementary internal angles of a triangle, to measure the vertical position of the Moon above the horizon. (The use of this instrument will be explained in Chapter 9.)

Instruments were to be of incalculable importance for the future development of science, for what lies implicit within them is the idea that precision is essential when it comes to an exact understanding of nature, and that the unaided human senses are, quite simply, too approximate to reach the heart of things. Instead, human vision needs *sights*, made of slits or pinholes, combined with exactly divided circles, if we really aspire to an accurate description of the night sky, for only then will the data be good enough for us to model nature and obtain sufficiently consistent results that can make theory match fact.

In addition to exact instruments for observation, the Greeks went on to develop instruments that could model planetary movements by way of machinery, for purposes of teaching or calculation. Ptolemy's *Almagest*, for example, describes an early form of astrolabe, made of brass, which could be used in this way. But most breathtaking of all is that bronze device brought up by sponge divers from a ship that sank in *c*.80 BC off the Greek island of Antikythera. This Antikythera Mechanism – now preserved in the National Museum in Athens – is nothing less than a *geared* calculating machine that was probably used for computing the positions of the planets. And while, around 44 BC, Marcus Tullius Cicero mentions that ingenious astronomical devices were manufactured in metal on the island of Rhodes (not far from

Antikythera), the Mechanism in itself is utterly unique in the ancient world, and nothing similar has survived. Yet its very existence, amongst the cargo of a trading vessel, opens our eyes to the fact that not only were the Greeks of 80 BC constructing intellectual models of the Universe that were based on the observed motions of the planets, but that they were going on to encapsulate these ideas in brass gear trains and portable calculating machines.

But when Hipparchus, Ptolemy, and other Greeks modelled the Universe by inventing complex mathematical configurations of epicycles to account for the planetary motions, did they believe that they were describing the 'real' Universe as it actually existed, or were they simply 'saving the appearances' by means of elegant mathematical schemes? For how, indeed, could solid, transparent crystalline spheres actually have these multiple trains of epicycles rotating within themselves? This tension between the modelling of nature so as to explain phenomena mathematically and the desire to know what nature is really like has, as we have seen, lain at the heart of science since long before Ptolemy. But it is a creative tension, with each style of thinking attracting different types of mind with their own unique insights.

Greek science was one of the most powerful intellectual tools to emerge from the ancient world, and without it modern science could never have been set upon its course. For while the Greeks were not the first people in antiquity to see nature as the product of a coherent design principle, they were the first to act upon that fact, and to view nature as something worthy of study in its own right. They were also the first to see, by logic and by mathematics, that great and consistent themes were present in nature, and that these could be linked with the eternal four elements, processes of becoming, or geometry. It was, furthermore, the Greeks who, in the whole history of mankind, first recognised that nature embodied profound truths that went beyond the obvious or the elemental, and that these truths could be observed, measured and modelled, and be found to possess predictive bases.

The Greeks ultimately realised that what made all these interconnections possible was their shared source in a transcendent and eternal order which linked Forms, number, mathematics, and their ability to be apprehended and explored by the human mind. And when this dynamic approach to knowledge came to be reinterpreted not so much as an aspect of the impersonal *logos* or *nous*, but in relation to the Jewish personal Creator God in the early Christian centuries, one of the most significant turning points in both the spiritual and intellectual history of the world had been reached.

SIX

From Jerusalem to Baghdad

Monotheism – the belief in one supreme all-powerful Creator God – lies at the heart of that coherent view of nature, and its connection with the human intellect, from which science developed. For in a culture which envisaged the creation of the world as a sort of mythological accident, in which competing deities produced storm, flood and famine, mankind always remained, in the last analysis, a victim of caprice. But when one unified mind or intellectual principle is seen as the source of all things, and that principle is perceived as being accessible to us through the divine gifts of reason, language and intelligence (even if that intelligence also teaches us the wisdom of obedience to a greater power), then we inevitably see the world differently. For nature's patterns need not be erratic, while the eternal truths of logic and geometry link our minds with that of the changeless Creator.

As we saw in the two preceding chapters, the Greeks developed the insights upon which their logic and science came to be based by the recognition of a rational agent – an impersonal *logos* or *nous* – that lay behind all things, and which was certainly older and nobler than

the fallible anthropomorphic gods and goddesses of Greek public religion. It was the Jewish scholar Philo of Alexandria in the early first
century AD, moreover, who (as we saw in Chapter 4) was especially
active in reconciling Judaism with classical Greek thought, arguing
that most of the central themes of Judaism lay embedded within it. In
this respect, Philo recognised the central significance of the Greek
rational principle of the *logos* to God's relationship with humanity. As
a monotheistic Jew, with a clear sense of the one personal Creator
God, Philo saw the *logos* as a sort of divine agent through whom God
the Father related to mankind, being, for instance, the intermediary
behind the burning bush through which God had spoken to Moses in
Genesis, and featuring in other Old Testament accounts of God's
contact with the human race. Indeed, Philo's attempt to reinterpret
Greek philosophy through the perceptions of monotheistic Judaism
was to have far-reaching consequences.

In the first century AD, however, a new force entered both
Jewish and Greek life: Christianity. For while Jesus Christ himself was
a Jew who had come to save the lost sheep of Israel, the power of his
teachings spread rapidly beyond the Galilean world of fishermen and
farmers after His crucifixion in Jerusalem around AD 30. Most especially, it was carried into the Greek-speaking world by an educated
Greek Jew who, under his original name of Saul of Tarsus, had
actively persecuted Christians. But after his miraculous conversion on
the Damascus road, Saul changed his name to Paul, and was to spend
the remaining thirty or so years of his life preaching Christianity in
Corinth, Ephesus, Athens, and other cities in the Greek world, before
residing and eventually dying in Rome around AD 64.

As an orthodox Jew, Paul saw God as the eternal being who had
created the entire cosmos out of nothing, as recorded in the Biblical
books of *Genesis*, *Psalms*, *Isaiah*, and the 'Apocryphal' Books of
Maccabees II and *Wisdom*. But Paul's Supreme Creator was not an
impersonal order-principle or mind like the Greek *logos* (with which

his education would almost certainly have acquainted him), but the stern yet loving Patriarch of Jewish theology. This Patriarch, moreover, had a constant and interactive conscious relationship with His creation which went back to Adam and Eve. In this creation, the undisputed apple of God's eye was the human race, made as it was by God in His very own image. What is more, this human race was given dominion over nature, in spite of that potential for disobedience which arose from its divine gift of free will, and which led to its fall, original sin, and need for redemption. Adam was invited by God to name all living things, to establish a human relationship with nature that was replete with tensions – sometimes benignly curatorial, sometimes nakedly exploitative, and sometimes finding nature rebounding in the form of snake-bite, disease or famine. On these topics, however, Philo and Paul would no doubt have been in agreement, and it would be interesting to know if Paul – whose ministry probably began a few years before Philo's death – was acquainted with Philo's ideas on God and the *logos*.

Yet where they would no doubt have radically differed is in the way in which Paul, after his conversion, came to see Christ as the incarnation of this one Supreme Creator God on Earth – as a being who not only had created the world and who existed outside normal time, but who also entered His own creation to save His own erring human race, and suffered crucifixion and resurrection on its behalf, before ascending back into a timeless heaven.

Irrespective of an individual's religious beliefs, it cannot be denied that Christianity brought radical new ideas into human thought. It elevated mankind to a status by which it was not merely in God's image, but by which God Himself was even willing to suffer on its behalf. For Paul argued that Christ was the prophesied redeemer of Israel, proclaimed by Isaiah and others, and at first took the new teaching to fellow Jews – those scattered, or 'diaspora', Jews who did not reside in Israel, but across the Graeco-Roman

Mediterranean world. And Christ was not just a prophet sent from God, but the Son of God, and hence the world-incarnated aspect of God Himself. When Christ returned to heaven after His resurrection, moreover, His Spirit or 'Holy Ghost' remained on Earth to be an active 'Comforter' to future generations. In this way, Christianity was unique among the monotheistic faiths in the respect that its one Creator God had three equal aspects: Father, Son and Holy Ghost, as expressed by theologians in the doctrine of the Trinity.

Yet Christianity possessed a universalism which from its very beginning transcended the traditions of any one ethnic or cultural group, for even Christ Himself had given comfort and hope to occupying Roman soldiers, Canaanite women, and other non-Jews. Likewise, Paul's own teachings were quickly taken up by non-Jewish Greeks and Romans, as were those of his fellow evangelists such as Peter, who went to Rome, and James, who went to Spain. Indeed, the wildfire speed at which the Christian faith swept across the ancient world, from Persia and Egypt to Gaul and Britain, is one of the most amazing facts of ancient history. Within three centuries, Christianity spread from Capernaum and the villages of Galilee into the very heart of the Roman imperial family, with the conversion of the Emperor Constantine and his mother Helena; and by AD 382 it was rapidly becoming the predominant religion of the Roman empire.

But what relevance has the origin and spread of Christianity to the history of science? I would argue that what Christianity did – in classical intellectual society – was to further engraft Jewish ideas of the one Creator God who had made the world *ex nihilo* (from nothing) onto Greek ideas of reason and geometry, to develop and consolidate the work of scholars such as Philo. For both Jewish Christians like Paul and their Gentile converts to Christianity from the classical pagan religions all shared the same creation story of the One God, Yahweh, who had framed the world from nothing and had entered into an enduring relationship with mankind. Potentially, this highly personal Creator God had far more interpretative potential for human thought than had an all-pervading but intangible *logos* or *nous*. Indeed, the early Christians were quick to see the potential that lay within both the Jewish *Yahweh* and the Greek *logos*, for at the beginning of St John's Gospel, where the creation of the world is expounded in the context of the Christian New Testament, God himself is equated with the Word, or *logos*: with that unmoved mover, from which all else comes, for 'In the beginning was the Word, and the Word was with God, and the Word was God. The same was in the beginning with God. All things were made by Him; and without Him was not any thing made that was made.' The author of St John's Gospel, moreover, firmly associated this *logos*, or Word, with God Incarnate in Jesus Christ. (Christianity's early grapplings with and final absorption and internalisation of Greek philosophy will be dealt with in the next chapter.)

The Christians, however, were not the first people to use monotheism, reason and geometry to create a culture in which science and astronomy had an especially dynamic creative presence. That, indeed, was to be the achievement of the Arabs.

By the early seventh century AD the Christian faith encircled the Mediterranean. In addition to Italy and northern Europe, Christianity had become the religion of the eastern Byzantine Empire, centred upon the city of Constantinople, and extending from Greece through modern Turkey and the Near East down into Egypt and northern Africa.

In AD 622, however, a merchant of Mecca named Mohammed had a divine vision. He was told to displace the various pagan gods worshipped by his fellow Arabs, and proclaim the One God in the already holy city of Mecca on the Red Sea. This God was to be found only through the discipline of repentance, prayer, and obedience – by 'submission to God', or Islam. Through a combination of electrifying religious zeal and military conquest, this new faith spread throughout the Arabian peninsula, into the Persian Gulf, north-west into Egypt, and along the north African coast into Spain – and it could have swept into northern Europe had not Charles Martel and his Frankish knights won a decisive victory at the battle of Poitiers in western France in AD 732, thereby preserving the independence of Christian Europe.

During the centuries after AD 622, a vast swathe of the Earth's surface – from India in the east to the south Saharan seaboard of Africa, into the west and north to Muscovy, Scandinavia and the Arctic Circle – was to be the domain of three religious groups who, in spite of their internecine strife, all worshipped the same One God. Whether He was called Yahweh, Allah, or was mystically united as Father, Son and Holy Spirit as the Incarnate God of the Christian Trinity, he had important attributes shared by Jews, Christians and Muslims alike. He was the only true God. He pre-existed time and space, and had made the cosmos from nothing. He had created the human race in his own image, and had a continuing, redemptive relationship with it. He was known through prophets and visionaries, and yet that wonderful design which he had stamped upon his creation at the beginning of time could also be appreciated through intellectual reflection, geometry, and Godly reason, as the human mind placed itself in harmony with the Mind in whose image it had been fashioned.

Jews, Christians and Muslims, moreover, were also 'the People of the Book' insofar as their respective faiths had a common written ancestry. Not only did they all share the same *Genesis* creation story, but many of the holy men and women from the Jewish Old Testament – Adam, Eve, Abraham, Sarah, Jacob, Joseph, Moses, and many others – passed into the Christian Bible and on into the Muslim Koran. And though the Muslims, in that uncompromising monotheism which they have always shared with the Jews, cannot accept that God ever stepped down from heaven to take human flesh in the form of Jesus Christ, the Son of God, they nonetheless revere Jesus as the last and greatest of the Jewish prophets.

In their character as the 'People of the Book', furthermore, the Jews, Christians and Muslims opened up a series of radical concepts in religious thought and its implications. Not only were they the loved, if disobedient, children of the One Creator God, but their relationship with him was seen as a thing of historical and literary record. Unlike the myths of the Egyptians, Babylonians, and even the Greeks, the spiritual genealogies to which the Jews, Christians and Muslims looked for guidance contained no equivocation. No-one was the father of someone in one myth, and their son or daughter in another. Indeed, the 'People of the Book' can be said to have had no creation or ancestral *mythologies* as such, insofar as all things came from one undisputed source – God – although what they had instead were what might be called a set of post-creation historical narratives of varying degrees of authority.

But the 'People of the Book' possessed another injunction: to study, ponder the world of God, and pray, rather than seek religious insight in animal sacrifice or divination. And while the Jews had combined study with sacrifice until, in AD 70, during the reign of the Emperor Vespasian, the Roman general Titus destroyed Solomon's Temple and its Holy of Holies, razed Jerusalem, and sent the Jews on their 1,900 years of wandering through the world, animal sacrifice never had a part in Christian religious practice, nor was it a part of Islam. Judaism, Christianity and Islam, therefore, became essentially contemplative faiths, in which one reached closest to God by mental exploration. Like the Pythagoreans and other Greek mystical cults, it was through the trained mind and disciplined spirit that one came nearest to glimpsing the divine, though the divine for which the 'People of the Book' sought was not a generalised principle of eternal order, but a sustaining personal relationship with the source of all things that penetrated and nourished every aspect of human nature.

It was for this reason that *science*, as that exploration of the divine as it was reflected in the natural world, had such a powerful appeal to the 'People of the Book'. And while Jewish communities scattered across the Mediterranean – such as that of Moses Maimonides in Spain, and of Levi Ben-Gerson in France – made original contributions to medicine, philosophy and astronomy, it was from the established territorial bases of Christendom and Islam that the true flowering of medieval science was to spring: from Damascus, Cairo, Toledo, Bologna, Paris and Oxford.

It was, moreover, Christian scholars who had carried Greek science into the Near East before the rise of Islam. Around AD 350, for instance, St Ephrem had founded a school at Edessa in Mesopotamia, and it was here that the writings of Aristotle, Ptolemy and other Greeks were translated into Syriac, which by that date was emerging as the literary language of the Near East. Christian scholars were also at work – not only spreading their faith, but also teaching the classical authors, at Jundishapur, to the north of the Persian Gulf.

When Islam leapt into being in the seventh century, however, it was not just from these sources that its first scholars drew their inspiration, but also from Indian, Persian and Nestorian Christian roots. Islam, therefore, absorbed many intellectual currents from both East and West (though apparently very little from China), and what was to be enduringly influential in the history of science was that system of numerical notation which we now call 'Arabic numbers'. From the ninth century onwards, however, it was the science of Ptolemy, Aristotle, and the other Greeks – by then translated into Arabic – which was to constitute the decisive influence on Arabic science. This was to take place very largely through the rise of the new capital of Baghdad after AD 762 and the establishment of its 'House of Wisdom' by the Abbasid Caliphs.

Yet why should the new religion of Islam be so open to Greek, Indian and other scientific ideas? Well, for one thing, the Koran itself taught that as God had made the world from nothing and in accordance with His design, then that world was a fitting object of human study. To explore the beauty and subtlety of the divine Creation, therefore, could be seen as a religious act.

This study, moreover, was useful on so many levels. A systematic study of plants, animals, minerals and other earthly objects was invaluable to medicine, farming, and technology; and the observation of the heavens was particularly important, for not only did the heavens comprise the most perfect part of God's creation, but they also promised to be very useful for practical purposes. The Muslims, like Jews and Christians, had an elaborate liturgical calendar through which to worship God, and this demanded a knowledge of new Moons, seasons and years, as Passover, Easter and Ramadan all have to be fixed from the Moon.

Direction-finding was also important. Christians needed to know the exact direction of east, for the alignment of churches and the burial of the dead (so that they would rise from their graves to see God face to face on the day of Resurrection), and Muslims needed to know the direction of Mecca from wherever an individual happened to be in the world, so that the five daily prayers could be said with proper reverence. And if you were in Spain, Saharan Timbuktu, Damascus or Calcutta, this presupposed a quite sophisticated knowledge of astronomy.

But there was another reason why science was so important to Muslims in exactly the same way that it was to be important to Christians, and this was cosmological. Cosmology was not merely concerned with the positions of lights in the sky, but about the vastly more important matter of proper order, design, and divine intention. Cosmology, therefore, was both practical and metaphysical, and at its heart lay number and geometry – which enables us to understand why the Arabs were so naturally attracted to Greek mathematics. The Muslims took with absolute seriousness the Jewish second commandment that 'Thou shalt not make unto thee any graven image', and interpreted it as a prohibition against making pictures of the human face or body (which by definition are reflections of the divine Being), and Islamic religious iconography could therefore contain no representation of the human form. Instead, the abstract perfection of complex geometry was used to encapsulate the glory and unity of the divine design. Mosques, illuminated manuscripts and other art works still testify to this aspiration, for Islamic cosmos and science (just like its Christian counterparts) were not simply about explanations for the behaviour of matter, but about ways of exploring God's ineffable mystery through those signs and wonders that He had left for mankind to follow. Science, therefore, was intimately related to theology, geometry to morality, and the quest for intellectual truth conjoined with an injunction to honest human dealing.

Both Islamic and Christian cosmology, therefore, were about ordering things into their proper physical and metaphysical places, for as with Plato and Aristotle, truth for them was also found in unity and hierarchy, and order reigned over chaos when everything was in its proper place. And as numerical relationships ran through classical Greek, Jewish and Christian cosmological explanatory schemata – forties, fours, threes, sixes, nines and so on – so they ran through those of Islam. There were, for instance, seven layers to the Muslim Universe, and seven regions of the world, as the contemplation of the physical elevated one to the divinely transcendent, for Pythagorean ideas about number permeated the intellectual life of all these cultures.

From an early stage in Muslim culture, and certainly by the tenth century AD, one sees a flowering of interconnected scientific disciplines across the Islamic world; and while their technical and conceptual vocabularies owed much to the Greeks and Romans, their metaphysical schema was that of the great Creator God of *Genesis*. Arabic medicine was based on the humoral physiology of Hippocrates and Galen, its natural history on Aristotle and Pliny, its physical cosmology on

Aristotle and Ptolemy, its mathematics on Euclid and the Indians, and its practical astronomy on Hipparchus and Ptolemy. This scientific culture was combined, moreover, with a broader blossoming of poetry, architecture, philosophy, and the arts of elegant living right across the Muslim world, from Baghdad, Damascus, Cairo and Basra in the east, to Cordova and Toledo in the west. It led to the resurgence of town planning (which had been largely in abeyance since the fall of the Roman world) in the eighth century AD, and the founding of public hospitals, academic libraries, and institutions of higher education.

It was these institutions, in fact, which came to be known by the generic name of Madrasahs, which were the ancestors of the later north European universities. Baghdad, Fez, Tunis and other Muslim cities came to have Madrasahs of distinction, and today one can still visit the Sultan Hassan Madrasah in Cairo, which was built during the 1350s AD. A Madrasah was first and foremost centred around a mosque. It would attract teachers of the Koran and Muslim theologians, often from different schools or traditions of Koranic exposition, while they in turn would attract students. In lectures and discussions, the nature of God would be explored, and insights and intellectual techniques borrowed from Plato or Aristotle would enter in. Young men would learn not simply how to worship God in a rigid way, but how to explore His glory with intelligence and ingenuity, through analogy, poetry and argument. These Madrasahs, moreover, would often acquire libraries containing Arabic translations of the Greek and Roman thinkers, to provide them with access to the intellectual treasures of antiquity.

In addition to lecture rooms and libraries, the leading Madrasahs also contained two other departments: hospitals and astronomical observatories. The Sultan Hassan Mosque in Cairo still has the now deserted narrow lanes of its hospital quarter. Here the sick would come for comfort and treatment, and teachers of medicine would do their equivalent of 'walking the wards' with their students. Diseases would be classified, their characteristic symptoms noted, prognoses made, and therapeutics based on Hippocrates, Galen, Diosciorides and others put into practice. Arabic medieval writers like Ibn-Sina (Avicenna, 980–1037) and Abu'l-Ibn-Rushd (Averroes, 1126–1198) compiled the first systematic case histories since the days of the Greeks, and their writings on treatment, drugs, and disease classification were faithfully copied out and distributed across the Muslim world. Indeed, their writings even came to be absorbed into Christian European medical practice, and by the fourteenth century had clearly become so familiar that a layman and poet like Geoffrey Chaucer could make a passing reference in his *Pardoner's Tale* to Avicenna's *Canon*, or rules of medical practice, which he would have known in a Latin translation.

It was, however, in astronomy that Islamic science made some of its most important and enduring contributions. Astronomy had a place in the Madrasah not only because the movements of the heavens formed a part of the study of the divine, but also because Arabic physicians saw the movements of the planets as playing an important part in the prognosis of disease. Health, after all, resulted when the four Greek humours of yellow bile, black bile, blood and phlegm were all in balance, and it was believed that these humours were susceptible to disturbance from the planets. Saturn, for instance, affected black bile, which in turn was the cause of depression, while Mars influenced yellow bile, which produced anger. An Arabic physician, therefore, would want to know where the planets stood in relationship to each other when a particular disease struck, and when forecasting its outcome, would interpret the patient's physical symptoms in accordance with the changing positions of the planets.

But Islamic astronomy went well beyond its possible services to medicine, for Arabic astronomers made the world's most detailed and sustained runs of celestial observations between the second and sixteenth centuries AD, with the three hundred years between 900 and 1200 as their 'golden age'. The creative centres of Islamic astronomy, moreover, tended (with a few significant exceptions) to move westwards over time, with their first burst of brilliance in Persia and Baghdad, with Al-Battani (*c*.850–929), Abd Al-Rahman Al-Sufi (903–986), Al-Biruni (970–1038), moving on to Cairo with Ibn-Yunis (late tenth century) and Alhazen (*c*.965–1039), and then to Spain with Arzachael, or Al-Zargali (died *c*.1087).

On a practical level, therefore, Arabic astronomy was about useful things like calendars and medical astronomy; but on a physical and mathematical level it was – like that of the Greeks – concerned with the abstract business of trying to explain the motions of the heavens. On the one hand this was driven by intellectual curiosity, although this curiosity always remained inextricably linked to an aspiration to enrich and deepen mankind's awe at God's creation. And having early on obtained the works of Ptolemy and immortalised his greatest under the title of *Almagest*, the Arabs began to rework Greek astronomy.

Ptolemy's *Almagest* is far more than an account of the heavens. As we saw in the last chapter, it incorporated the then surviving but subsequently lost works of many earlier Greek astronomers, together with long runs of observations, records of phenomena such as the precession of the equinoxes, and the dates of ancient and modern eclipses, which made it also an astronomical encyclopaedia. Ptolemy had forged all of this into a tightly argued geometrical scheme, each step of which progressed logically from the one before. The Arabs, like the Christians, found it all immensely appealing to the rational intellect, although it was the Arabs who first fully internalised Ptolemy within their intellectual culture, and learned how to think creatively within its theorems and propositions. But when they used many of the astronomical constants laid down by Ptolemy in AD 150, they found that they often did not quite work. So, was Ptolemy wrong in his details? Or had the heavens changed? The only way to find a reliable solution was to reobserve the heavens afresh, and obtain new data which could be carefully compared with that inherited from the Greeks. With Ptolemy's *Almagest* as a guide, backed up with the techniques for circle and angle division described in Euclid's *Elements*, the Muslim astronomers were able to build large shadow scales, quadrants, armillary spheres and other instruments with which to observe the sky anew for the first time in 700 years.

Abd Al-Rahman Al-Sufi produced a *Book of Constellations* which provided the basis of a revised star map which brought Ptolemy up to date. And from his time onwards, modern scientists, using large-radius and very accurate angle-measuring instruments, made meticulous observations of the angular positions of the stars and planets. Al-Battani obtained a much more exact figure for the obliquity of the ecliptic – the angle at which the Sun's path, or ecliptic, intersects with the celestial equator (as described in Chapter 5). Then, in the tenth century the Persian Abu'l-Wafa Al-Buzjani discovered a new, tiny irregularity in the Moon's orbit around the Earth, which we now call evection. It was independently discovered by Tycho Brahe in Denmark in the sixteenth century, and not until European scholars started to work on Arabic sources in the nineteenth century was it realised that Abu'l-Wafa had already identified it more than 600 years before Tycho.

Many observatories that did outstanding work in practical astronomy came into being in the Muslim world during those centuries which we refer to as the 'middle ages', and though the golden age of Islam's most original scientific culture drew to a close in the thirteenth century AD, its best-equipped observatories, ironically enough, were built after 1200. In 1259, for instance, Persian astronomy enjoyed a revival when a major new observatory was established at Maraghah, but without doubt the finest of all Arabic research observatories was that built by the Mongol prince Ulugh Beg (1394–1449) at Samarkand in 1424. Ulugh Beg was a Muslim of Tartar descent, in the line of Tamerlane, and ruled Samarkand, which stood well to the east of the heartland of Arabic culture, in the province of Maveranakr. But the Prince, in addition to living in rich oriental splendour, was an intellectual. He founded a Madrasah, but instead of simply handing things over to a staff of professional academics, he led the way personally when it came to astronomical research.

At Samarkand, he designed and built what is probably still the world's largest-ever sextant. Its 60-degree scale (hence *sextant*, from the Latin for ⅙ of a 360-degree circle) had a radius of 130 feet, and was oriented to stand in the vertical facing due south in the plane of the meridian. The 60-degree scale was built as an integral part of the great, drum-shaped 100-foot-diameter masonry building, complete with observing galleries, computing rooms and teaching rooms which constituted Ulugh Beg's great observatory. At a radius of 130 feet, each degree on the quadrant's scale was 27 inches long, each arcminute almost half an inch, and 10 arcseconds an easily discernible angle.

To use this vast instrument, Ulugh Beg would sit on a curved flight of stairs that ran behind the sextant's 60-degree divided scale, and look through a special sighting hole in a wooden board. From there he would look on through a second sighting hole that was aligned with it, 130 feet away, at the geometrical centre of the sextant. It was, in fact, rather similar to a gigantic rifle sight. Then, as his line of sight passed through the two holes, it proceeded on to the clear sky, so that he could observe whichever objects were currently due south within a compass of 60 degrees in the vertical.

No astronomer before Ulugh Beg had built an instrument which could read down to such tiny angles, and it was in his driving concern for greater precision that we see not only what drove him but what also drove many other Arab astronomers. He wanted to know the *exact* highest and lowest points of the Sun in the sky at the summer and winter solstices, for this was the key which could unlock many other details about the Sun's (or Earth's) orbit. In consequence, Ulugh Beg was able to use the great accuracy of his sextant to determine the length of the solar year to within a minute of time of the figure which we accept today – a vital piece of information when it comes to understanding planetary movements. In addition, he wanted to know the exact quantity of the precession of the equinoxes, and using his great stone sextant in conjunction with other instruments, he also made a map of the positions of 1,018 stars. Indeed, Ulugh Beg was one of the greatest astronomical observers of the pre-telescopic age, and the accuracy of his observations, when they became available in the West in Latin translations in 1665, provided crucial comparative data for John Flamsteed and later European astronomers.

Ulugh Beg's observatory, Samarkand.

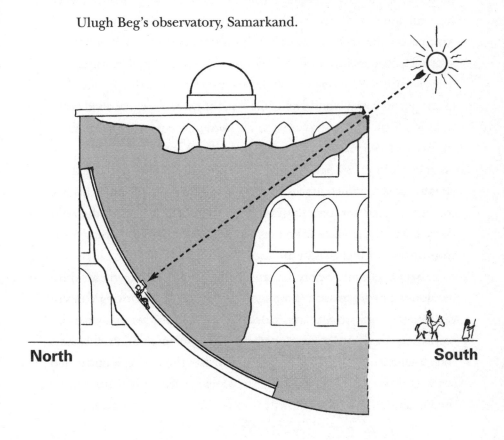

North   **South**

Yet tragically, Ulugh Beg's great contributions to science and learning were cut short when he was murdered in a political coup engineered by one of his own sons. His Great Observatory was torn down, and was lost for centuries – until 1908, when Russian archaeologists excavated the pile of rubble which was believed to mark its location. The Russians unearthed the lower degrees of the stone sextant, running as they did below ground in a deep meridional trench, and from its surviving radial curves and some remaining degree markers they were able to calculate the sextant's and the rest of the observatory's exact dimensions. The ruins of Ulugh Beg's observatory are now preserved as a major archaeological site in Uzbekistan, although perhaps the best known and most beautifully preserved Islamic observatories are those built by Jai Singh (1686–1743) at Delhi, Jaipur, Ujjain and Benares, with their magnificent quadrants and shadow dials in polychrome marble, each piece of which is an architectural masterpiece in its own right. And while these Indian instruments are of very late construction, dating from the eighteenth century AD, they nonetheless convey a sense of the monumental proportions and wonderful equipment of those great scientific institutions which, for the first time since antiquity, were used to compile the first detailed records of celestial movements.

In addition to performing observations of hitherto unprecedented accuracy, Islamic astronomers made major innovations in the organisation of the resulting astronomical data. Accurate observations, after all, are of only limited value if they cannot be placed in a context in which they may be used for comparison and analysis. In consequence, the Arabs invented what they called the Zij – a form of handbook or tabular arrangement of data which made it easier to spot consistencies and select data for mathematical analyses. This use of tabulation was further developed into an astronomical handbook, often calculated to be suitable for a particular 'climate' or location on the Earth's surface, and which in Arabic was called an Al-Manunkh – or almanack, as we now call it.

The most enduring astronomical innovation to come out of the Arabic world was that instrument called the astrolabe, which was second only in influence, perhaps, to the almanack. Like the quadrant and Ptolemy's Rulers, the Arabs did not invent the astrolabe without a precedent, but took its basic principles from Ptolemy. However, by the tenth century they had transformed it into a sophisticated array of rotating brass plates which, in effect, constituted an early form of personal computer.

Astrolabes could be anything from three to fifteen inches in diameter, being made up of a set of (usually five) circular brass plates that nestled inside a larger circular base plate. Each side of the five plates would be engraved with a set of latitude and longitude lines, calculated to represent the angle of the north pole star from a given place on the Earth's surface, such as Mecca, Cairo or Damascus. A star map, made in the form of a brass filigree, rotated above the plate that was appropriate for a given location, so that it was a device which made it possible to produce a perfect simulation of the rising and setting of the stars for that location. Once star risings could be replicated, the instrument could be used to calculate all manner of things: sunrise and sunset, the time of day or night, the future phases of the Moon, the rising of zodiacal signs, and much else besides. Indeed, the astrolabe was not only to become widely used across the Muslim world – rather like personal computers are used worldwide today – but spread beyond it into northern Europe and India. Until western Europeans developed logarithms in the seventeenth century, the astrolabe remained the quickest and most accurate way of calculating the coordinates of what astronomers call 'spherical triangles' – large, curved expanses in the night sky, such as the geometrical relationship between individual stars in the zodiac and a planet. And perhaps no finer testimony to the universality of the astrolabe is to be found than in the many hundreds of instruments that still survive in museum collections around the world – most notably, in the Museum of the

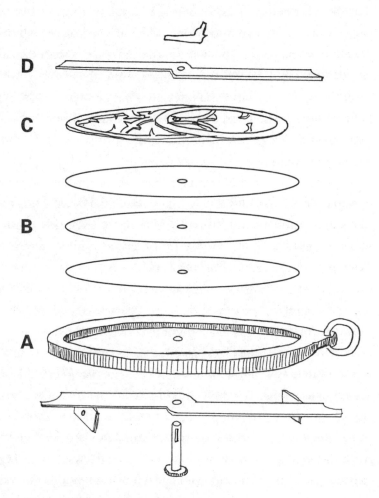

The parts of an astrolabe. The
thick, hollowed-out base-plate, 'A',
houses the 'climate' plates, 'B',
onto which the sky-projections for
Cairo, Baghdad, and so on, are
drawn. Above them is positioned
the 'rete' or filigree star map 'C',
along with the 'alidade' ruler, 'D'.
The whole instrument is locked
together with a brass rivet and peg.

History of Science, Oxford. Some of these astrolabes (dated not, of course, from Christ's birth, but from Mohammed's journey, or 'Hegira', of AD 622), are more than 1,000 years old.

Ultimately, however, Arabic astronomy still found itself wrestling with the same set of problems with which the Greeks had struggled a millennium earlier. These problems included the following. How did one reconcile a theoretical prerequisite for perfectly circular planetary orbits and uniform orbital velocities with *observed* variations in both? Did the resulting gear-wheel chains of spheres, epicycles and eccentric motions really describe the Universe as God had made it, or were these but mathematical inventions that 'saved the phenomena' but did not describe reality? And how, moreover, did one reconcile the scientific approaches of Ptolemy and Aristotle, one of whom sought solutions via mathematical elegance, and the other through the observed interaction of vital forces within nature? One might argue, however, that these problems were more pressing for the Arabs than they had been for the Greeks; for the Muslim scientists, with their acute awareness of the active presence of a singular Creator God, were more concerned with discovering the actual truths within nature. If one's perception of the Creator was as a generalised rational force – as it had been for the Greeks – then one could perhaps work more easily with different philosophical models of reality; but if one's God was personal and immanent, and you had a holy text which described how God had brought everything into being, then it seemed that there must be only one correct way of describing the world, and that it was the scientist's job to determine the actual truth.

It is also possible to argue that by the time Arabic science came to its full flower, the problems were becoming deeper rather than easier, for the long runs of superbly accurate astronomical observations, arranged into Zij and Al-Manunkh tables, and analysed by superior mathematical techniques, only highlighted more and more discrepancies from the constants established by the Greeks. This led, on the one hand, to attempts to 'perfect' Ptolemy's *Almagest* in the light of new observational and mathematical data, and on the other hand, to a burgeoning of the Aristotelian *qualitative* approach to science, with its emphasis upon nature's *properties* – such as light, motion, comparative observations, and study of equilibrium systems – rather than upon abstract mathematical modelling. In many ways one might argue that these two approaches to science reached their respective peaks of development in the astronomer Nasir Al-Din Al-Tusi, and in the optical physicist Ibn-Al-Haytham, better known in the West as Alhazen.

Nasir Al-Din Al-Tusi (1201–1274), of the Maraghah Observatory, Persia, worked within that tradition of Arabic astronomy which absorbed observations and mathematical models developed from Arabic predecessors to try to 'perfect' Ptolemaic astronomy within the light of the best and latest knowledge. But Al-Tusi (as he is generally known) was one of the greatest 'celestial mechanicians' of the Islamic world, and in his attempts to explain the observed 'inequalities' of the motions of the planets within the context of the Greek requirements for uniform circular motions, he devised some remarkable geometrical models incorporating epicycles and eccentric circles. Some of these models, moreover, incorporated circles with centres which seemed to slide along an axis – at a perfectly uniform rate, of course – so that the resulting planets were describing what were, in effect, *elliptical* orbits while moving around the Earth.

By 1250, therefore, Islamic astronomers seemed to be willing to stand on their heads in their attempts to find ways of explaining the apparently irregular motions of the planets, such as their retrograde loopings, within the rules of uniformity laid down by the Greeks, and their solutions became ever more abstract. Yet had that all-powerful and perfect Creator God who had first set the planets spinning in space, and to whom truth was elegant and simple, *really* used this fantastic over-complication in His design? Ultimately,

Arabic astronomy became bogged down in the proliferation of its increasingly complex explanations of the simple. And yet these Arabic observations and planetary models could well have been of use to the Polish astronomer Nicholas Copernicus after 1510, and to the German Johannes Kepler after 1608, in their development of a heliocentric Universe (as we shall see in Chapter 10). For while we cannot be sure that Copernicus and his successors were acquainted with the celestial mechanics of Al-Tusi, it cannot be denied that Copernicus explained certain planetary irregularities by means of geometrical and physical devices that were very similar to those of Al-Tusi.

Ibn-Al-Haytham – or, as he became known in the West under his Latin name, Alhazen (*c.*965–1039) – was just as fascinated by celestial geometry as were his fellow scientists, but his enduring contributions to Arabic science were not so much concerned with planetary motion as with the nature of light, which he saw as coming directly from God. It was Ptolemy's optical writings, and especially Aristotle's *Meteorologica*, which dealt with the physical characteristics of the Earth's atmosphere, which were influential in his thought.

Alhazen, who was a native of Basra, on the Persian Gulf, spent much of his working life in Cairo, where he seems to have practised as a physician, in addition to pursuing his wider scientific interests. In his studies of light, he dissected the eyes of freshly slaughtered oxen and other large animals to observe their structures, though as the Koran forbade the dissection of human corpses (as humans were made in the image of God), he might never cut open a human eye. He noted that the lens in an animal eye was a transparent piece of gristle, and that it changed focus by changing shape. He also subsequently identified the iris, the cornea, and other parts of the eye, and traced the path of the optic nerve to the brain. Alhazen recorded that while one could experiment on freshly dead eyes, after a while the lens became cloudy, and the other parts of the eye lost their shape.

Light fascinated the Arabs, as it had fascinated the Greeks and would fascinate the Christians – not only because it flooded the world with radiance, but also because it seemed to be the only thing that came from the realms of divine space into the world of mortals. Light, moreover, could not be bottled, weighed, or measured, and seemed instantaneous in its transmission, although it still reacted with earthly things. White light obviously changed to red at sunset, decomposed into six colours when it was passed through a glass vessel filled with water, and could be reflected and bent. Indeed, it is in these studies that we find the beginnings of the science of optics, and Alhazen, perhaps more than any other Muslim, Jewish or Christian scientist of the medieval age, laid the foundations of that science.

Optics, like astronomy, is at heart a mathematical discipline, for certain colours or images always appear in the same places when the light shines from a particular angle through a block of glass or through a cloud. And as Alhazen was to demonstrate brilliantly, optics was not just a mathematical science, but also an *experimental* one, in which one could devise particular things, such as refraction and reflection, for the light to do, from which one could attempt to define its physical characteristics.

One major optical truth which Alhazen almost certainly established was how the eye perceives. Perception, after all, lies at the heart of all scientific knowledge, and without the perceiving eye, we can neither study plants and animals, nor measure the stars in their courses. Since antiquity, there had been uncertainty as to whether light entered the eye from outside to produce a focused image that impinged upon our consciousness, or whether the eye was somehow 'sensitive', and had something like invisible feelers that radiated from it and 'sensed' the world. Alhazen's researches inclined him towards an 'intramission' (light entering the eye) model of perception (which is, of course, correct), as opposed to the 'extramission' (sensitive feelers) model.

He also performed ground-breaking experiments that investigated the refraction or bending of light as it passed through glass vessels containing water, along with glass cylinders, spheres, and similar rudimentary lenses. His work on the optical characteristics of plain, spherical and parabolic mirrors was also fundamental to the subsequent establishment of the laws of reflection and image-formation. And while Alhazen did not invent the pinhole camera (whereby one could produce inverted pictures of the natural world by admitting light through a pinhole into a darkened room or box), he *did* study its image-forming properties.

Yet perhaps Alhazen's most original optical researches were concerned with the investigation of the passage of sunlight through the Earth's atmosphere, and he performed important work in the optics of the rainbow. He suspended a glass sphere in a darkened room, and passed a ray of sunlight through it, which enabled him to trace the refractions and reflections which produce the colours and directions of what physicists call the primary and secondary rainbows. In particular, Alhazen realised that the Earth's atmosphere, far from being pure and optically neutral, produces all sorts of distortions, especially when one looks at objects close to the horizon, where the air is at its densest. It is this dense air, for instance, which makes the Sun look larger, fiery red, and more egg-shaped as it rises and sets, whereas when seen high in the sky through cloud it seems smaller, white, and perfectly circular.

But what causes the sky to become light in the morning, long before the Sun appears above the horizon, and to retain some light for perhaps an hour after sunset? Alhazen realised that the air scatters and bends light, and from careful angular measurements, probably made with an astrolabe, he was able to discover that when the first light of dawn appeared directly overhead, the Sun was still 19 degrees below the horizon. The same also applied after sunset. This happens because the Earth's atmosphere acts like a large lens, and

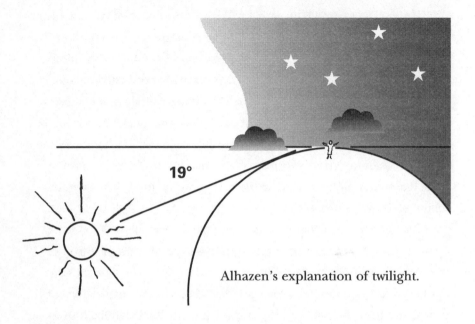

19°

Alhazen's explanation of twilight.

refracts the straight solar rays, to scatter them in the atmosphere above a location still in darkness.

The approach to natural knowledge displayed by Alhazen was rather different in its assumptions from that of celestial mechanicians such as Nasir Al-Din Al-Tusi. Alhazen's approach is less concerned with mathematical modelling and more with observing and experimenting: much more Aristotelian than Ptolemaic, in fact. One could justifiably argue that Alhazen was medieval Islam's greatest physicist, and the Latin translations of his works – especially his *Optical Thesaurus* – were to be of the utmost importance in encouraging thirteenth-century figures such as the Englishman Roger Bacon to take up optical physics where Alhazen had left it (as we shall see in Chapter 8).

Within the bigger picture of the global development of astronomy and its related sciences, however, we must not lose sight of the fact that the Greeks and the early Arabs were not the first people to show a rational concern with the heavens. The cultures of the Indian sub-continent developed sophisticated astronomical and especially mathematical systems, as did the pyramid-building Aztecs of central America, although it was the Chinese who produced the longest and most sustained run of astronomical observations. To follow the scientific histories of these cultures would, however, require a wholly separate book, for in many respects they had grown from different philosophical and theological bases.

The classical Chinese, for instance, had by the third century AD developed a sophisticated cosmology based on the spherical Earth and sky, which had some parallels to the ideas of Eudoxus and certain other Greeks. Yet devising schemes of interconnecting celestial spheres to explain planetary retrograde loops was less important in Chinese cosmology than it was in the West. It could be argued that Buddhism – with its emphasis upon infinity and vastness rather than upon mechanism – had something to do with this. Similarly, Taoism and Buddhism were not religions that focused upon the deliberate actions of one single Creator God, so that the Greek, Jewish, Christian and Islamic concerns with a single unified design were less relevant. The Chinese, moreover, having at an early stage in their scientific history obtained an accurate value for the length of the solar year, at 365¼ days, were left with an awkward number into which to divide the zodiac. One can, after all, do wonders of division with 360, but 365¼ is hopeless!

It was perhaps for these reasons that China, in spite of its sustained accuracy in the practical business of astronomical investigation, never developed systems of coordinate geometry which were remotely parallel to those of Euclid or Apollonius. And without such systems of geometry, a truly predictive basis for the analysis of planetary motion is not possible. This may also have been why Chinese astronomical thought had so little influence upon the Arabs, in spite of the fact that the Arabs traded with the Chinese via Karakorum and the Silk Road across central Asia.

It was the Arabs, therefore, who were the first to inherit the scientific mantle of the Greeks, who took Aristotle, Ptolemy and all the rest of them greatly to heart, and who combined them with new spiritual insights derived from monotheism. From that model for the unity of all knowledge which they took from the Greeks, and from Aristotle in particular, the Arabs made significant advances not only in astronomy, but in geometry, pure mathematics, clinical medicine, botany, mineralogy, chemistry, and many other branches of learning.

The Arabs, moreover, were great namers of things and coiners of technical terms, and many of these slipped so effortlessly into medieval Latin and on into the modern languages of Europe – via the Latin translations of Arabic scientific books – that we rarely think of their true sources. They include star names such as *Aldebaran*, *Altair* and *Betelgeuse*; astronomical terms such as *nadir* and *zenith*; and mathematical words such as *algebra*, *algorithm*, *cipher* and *zero* (the mathematical concept for nothing, first invented by the Arabs). Then there are chemical terms like *alembic*, *alkali*, *borax*, *elixir* and *talc*; musical instruments such as the *lute* and *rebeck*; and such familiar chemical substances as *sherbet*, *coffee* and, of course, *alcohol*. Indeed, if ever one encounters a scientific word with the prefix *al-*, it will as likely as not denote an Arabic origin (although there are exceptions: *altitude* derives from the classical Latin *altus* – 'high' or 'deep').

Yet while the Arabs were making prodigious strides in the assimilation of Greek science into their own burgeoning culture, and undoubtedly led the way scientifically until the twelfth and thirteenth centuries AD, it is wrong to think that nothing of any value was happening in Christian Europe.

SEVEN

The City of God

After about 1650, one of the creedal assumptions of the Enlightenment and post-Enlightenment worlds was that between the glories of classical antiquity and the supposed 'rebirth' of learning in their own time, the intervening 'middle ages' had been 'dark' and given over to barbarisms and superstitions of the grossest character. All of this, needless to say, had been perpetrated with the blessing of a repressive, witch-burning Christian Church. Indeed, this obsession with the 'darkness' of medieval Christian Europe has become so firmly embedded in the popular consciousness of the modern world that journalists routinely use the word 'medieval' as a synonym for anything that is uncouth, and no big box office film dealing with the period is now deemed complete without its quota of filthy debauched priests, torture, cruelty, excrement, plague, and, of course, lots of darkness.

But this visualisation of medieval Europe is itself a piece of modern mythology that stands upon foundations of prejudice and fantasy that are no less historically outrageous than the more recent notion that aliens from outer space built the pyramids. The concept of the 'dark middle ages' was by and large the invention of men with an agenda. Men like John Locke, Voltaire, David Hume, Denis Diderot, Etienne Bonnot de Condillac, and others who were self-consciously 'modern', saw the recent scientific revolution as a liberating time for mankind, and liked to juxtapose their own perceived intellectual freedom and lack of serious religious belief with the supposed tyranny of the Christian Church which had preceded it. Yet, like all attempts to rewrite history from a particular ideological stance, what it produced was a travesty of the past, in which the inevitable disasters that took place across a span of a thousand years were elevated to the status of characteristic tendencies, and the real enduring contributions of Christian Europe very much ignored.

But what was it about the Christian faith which, within the space of little more than three centuries, took it (as we saw in the previous chapter) from the villages of Galilee to become the principal religion of the Roman empire? Its obvious and most powerful strength was the promise of salvation and eternal life which it offered to all men and women. For though it began within Judaism, Christianity was not an ethnic religion. Birth or patrilineal descent was irrelevant. In the *Acts of the Apostles*, at around AD 50–70, in fact, one encounters Jews, Greeks, Romans, Persians, Arabs and Ethiopians who had accepted or were actively interested in the Christian faith, along with a social spectrum that included beggars, senior court officials, officers in the Roman army, Greek intellectuals, and women. Access to the Christian faith, moreover, did not require any gifts of intellect or any educational skills. No secret mysteries or abstruse philosophical systems had to be mastered, and while St Paul's letters to the early Church communities and the four Gospels were early written works which narrated the life and teachings of Christ, personal literacy skills were not necessary for the ordinary Christian. The only requisites were sincere repentance, loving one's neighbour, and faith in the Resurrection through Christ.

In these respects, Christianity constituted one of the most radical turning points in the history of ideas, for no previous set of religious teachings or philosophical system had been so completely open in its ethnic, intellectual, gender or cultural appeal. Quite literally, anyone could become a Christian, and between AD 30 and AD 330, tens of millions of people did so. But why?

In the ancient world, religion in all of its varieties played a vastly more important role than it does in most people's lives today, irrespective of whether one worshipped Marduk or Ra, offered sacrifices to Yahweh, consulted the oracle at Delphi, or sought the *nous* through Pythagorean number mysteries. Religion and the gods not only explained chaos and order in the world, but also a whole range of spiritual phenomena. All the religions of antiquity teem with accounts of visions, ghosts, prophecies, predictive dreams, and other incidents which would once have been universally recognised as 'spiritual' but which modern psychology tends to interpret as trauma-induced illusion or as wish-fulfilment.

In the past, the human soul was generally held to be the somewhat loosely tethered vital essence of a person – loosely tethered, that is, in the respect that it could leave the body and wander off on its own, especially at night or during bouts of fever, after which its peregrinatory experiences were recollected as dreams or visions. What happened to this soul after a person's bodily death was a subject of intense interest in antiquity. The Egyptians believed that it lived on beyond death to experience an eternal lifestyle similar to that which it had lived on Earth, serving or being served. The Greeks and Romans generally thought that it descended into the subterranean region of Hell, or Hades: a dingy, aimless and immensely boring place where the dead were but shadows of their former selves, and which was ruled over by Pluto, king of the underworld. In his *Aeneid*, the Roman poet Virgil speaks of the gloomy, pestiferous and birdless Lake Avernus as the entrance to Hell. The later Jews also had similar ideas about a dark place called Sheol, which was the forbidding 'Pit' referred to in the *Psalms*. Others favoured ideas of transmigratory souls that encapsulated the eternal life principle and migrated from one body to another after death. Socrates clearly thought along these lines, and expounded them at length in the *Phaedo*.

The future fate of the soul, therefore, was a subject of serious consideration in antiquity, and what Christianity had to offer was something far more delightful than anything else available within that spiritual culture. For it offered eternal life to each individually resurrected body, made whole and restored from the ravages of age, sickness, deformity and sin, to sit with Christ at the celestial banquet. By definition, heaven's joys would never end, nor, as they were in the realms of blessedness and divine perfection, ever grow wearisome or boring. To receive these joys, one had only to sincerely repent of one's sins, and accept Christ as saviour. If, on the other hand, one consciously and deliberately rejected Christ – as beings possessing free will were at liberty to do – then one could only expect to enter that place of fire, brimstone, wailing and gnashing of teeth in which there was only eternal torment: hell. Even so, as Christ's teachings were first and foremost about forgiveness and salvation, rather than about damnation, certain theologians, such as Origen and Clement of Alexandria, argued that even those condemned to hell would eventually be reconciled to Christ and be saved.

Indeed, it was this extraordinary universalism which made Christianity so appealing, especially to the non-Jewish Greeks and Romans, by the end of the first century AD. In that world of clearly-defined power structures, Christianity was dangerously egalitarian and potentially subversive. For Jesus Christ himself had been no respecter of worldly social rank, being willing to rebuke high priests and rulers while blessing and identifying with society's outcasts. What mattered was not one's worldly status but only one's willingness to repent and to be 'born again' in Christ. Everyone, therefore, stood on equal terms in the sight of God.

It was all the more remarkable, therefore, that Christianity spread so rapidly in a Roman world which had grown to greatness by a national policy and a public religion that seemed to embody the very antithesis of Christian submission. Roman public religion was about power, victory, and fertility, and the Roman governing classes saw themselves as possessing a divine mandate from their gods to overcome barbarians and to establish the *pax Romana* across the world. And that, of course, meant peace under *Roman* terms.

The Roman gods were in many respects planetary deities, and one can see their relationship with Greek, Babylonian, and other gods. Sol Invictus – the invincible Sun, with his streaming rays – is perhaps the deity depicted in the Sun-like 'Gorgon's Head' circular carving surviving in the Roman Museum, Bath, while he may also have carried resemblances to the Babylonian Sun-god Shamash. Jupiter, the leading god in the Roman pantheon, had clear parallels with the Greek Zeus; Venus, the Roman goddess of fertility and love, with the Greek Aphrodite, Syrian Asarte or Ashtaroth, and Egyptian Hathor; while Mercury, fast-moving messenger of the gods, may be equated with the Greek Hermes. The Roman Luna likewise ruled the Moon in very much the same way as her Greek equivalent Selene had done. Blood-red Mars was the Roman god of war, while dull, slow-moving Saturn embodied old age and the passage of time, very much as he had for the Greeks. Indeed, these divine genealogies and attributes were determined by Roman and later Greek writers such as Cicero, Virgil and Plutarch in the centuries before and after the birth of Christ, and provide modern scholars with interesting evidence revealing how the Romans perceived the ancestry of the planetary gods across that Mediterranean and Near-Eastern world which they were coming to conquer. Christianity, however, came from a very different direction, and brought a very different message.

It would be wrong, however, to view the spread of Christianity across the Roman world, and especially within Italy, as one of unopposed triumph. For there were indeed persecutions and purges, often initiated by emperors or senior public officials who believed, quite rightly, that Christianity was teaching a very different set of virtues from those upon which Rome's greatness had been built. Yet the endless diet of Christian flesh which Hollywood epic-makers would have us believe had become the staple food of the Roman Colosseum lions was not an historical reality. Emperors like Nero in AD 64, and Diocletian around AD 304, certainly turned their spasmodic fury upon Christians, and it was in the purges of AD 64 that St Peter – Christ's closest disciple – was crucified upside-down in Rome, and St Paul probably also perished, but Christian persecution was by no means sustained. One of the reasons for this lay in the remarkable religious tolerance displayed by the Roman state; for generally speaking, if one was willing to be a good citizen, then one was relatively free to worship one's chosen gods – most of whom, after all, were seen as being related to, or aspects of, the Roman gods.

But where Christians, and the Jews before them, had seemingly invited persecution from the Roman authorities was in their blunt refusal to partake in Roman public religious life. Belief in the one Creator God of the Jewish Law, or in His incarnation on Earth as Jesus Christ, was unequivocal in its utter denial of divinity to Jupiter, Mars, Venus and the rest of the Roman pantheon. These gods were, quite simply, non-existent idols, and an abomination to God Himself. To the highly civic Romans, it was this lack of willingness to countenance a major component of the state's corporate life which was a particular source of annoyance when it came to the matter of Jews and Christians. Indeed, it says something about the remarkable tolerance of Roman society that Christianity was able to spread so fast and so far; a tolerance, moreover, that Christians were unwilling to show for either the beliefs or the artworks of their non-Christian fellow citizens once they had gained their ascendancy.

In the three centuries that followed Christ's earthly ministry, many of the basic structures of the new religion were established. Subsequent Christian historians, for instance, were to date the origins of the papacy from this time. The earthly Christ, for example, had told Peter that upon him the Church – or fellowship of believers – would be built after Christ Himself had returned to heaven. This was seen as making Peter the flesh-and-blood father of the Church: *pater* or *papa* in Latin – hence Pope. And as Peter was the disciple closest to Christ and received Christ's commission to evangelise the world, when Peter himself died this commission, or 'charisma', was said to have passed on to a successor, Linus, and from Linus on through an historical succession of heads of the Church down to our present-day Pope John Paul II. It was to Peter, now sanctified, that the premier church in Christendom – St Peter's, Rome – was built and consecrated. And from the Popes were ordained all of the bishops, priests and deacons who made up the clergy, in what was believed to be a pipeline of divine authority.

It was also during these early centuries that scholars began to assemble those letters, historical narratives and visionary experiences which formed the Gospels and other documents of what would become the New Testament. They also faced an editorial task, for it is clear that in the strongly polytheistic world of antiquity, some pagan and semi-pagan religious teachers had borrowed parts of Christianity to engraft onto their own particular teachings, and it was necessary to sift through what was a truly substantial corpus of Christ-related documents in circulation, to separate the authentically Christian sources from those which mixed Christian components with pagan teachings or else were deemed unorthodox on other grounds. It was also necessary for scholars to evaluate the authority of a range of older Jewish messianic texts which prophesied the coming of the Saviour of Israel, and arrange them to form the Old Testament which, when combined with the Christian texts in the New Testament, constituted what was called in Latin the *Biblia Sacra*, or, as it became in English, the Holy Book or Bible. By the early fourth century AD much of this work was already accomplished, and after the early bishops and Church leaders had met for the great Council at Nicaea, in Greece, in AD 325, the basic Christian creeds and doctrines were essentially in place, and the Christians had a 'Book' – the Bible – by which to define their spiritual identity and guide their conduct. And then St Jerome (*c.*342–420) took these predominantly Greek texts and produced from them a unified translation in the ordinary or 'vulgar' tongue of the Romans: Latin. This definitive Vulgate Bible would survive unchanged for 1,000 years, until Reformation scholars translated it into German, English, and the other vernacular languages of Europe.

But how did these teachings penetrate to the very heart of the Roman empire, and go on to change the course of the history of the Western world and beyond? Well, the Emperor Constantine (*c.*280–337), who had been proclaimed Emperor at York in 306, had a vision of the Cross in the sky as he fought a decisive battle against Maxentius, his imperial rival, at the Milvian Bridge in the approaches to Rome in AD 312. He took this to mean that Christ was signifying his intention to work through the Roman Empire for the salvation of the world. Yet though he was sympathetic to Christianity, and had a mother, Helena, who was a devout collector of holy relics (in 326 she sent a commission to Jerusalem to search for them, and it allegedly found the *titulus*, or headboard, that Pontius Pilate had ordered to be nailed to Christ's cross at the crucifixion), it was only at the end of his life that Constantine accepted baptism and became a Christian. Even so, a Roman emperor had died a Christian. In AD 382 Christianity effectively became the official religion of the Empire, following the removal of pagan altars from the senate and other official places – much to the dismay of the worshippers of the old gods.

From this time, moreover, other things began to happen in the Roman state which would have seemed unimaginable a couple of centuries earlier. For example, in 408, pagan temples lost their right to hold property; although long before that happened, Christian churches were being erected and consecrated in Rome. Around 326, the Empress Helena built the Church of Santa Croce (Holy Cross) to house her Jerusalem relics, while her son Constantine was responsible for establishing St Peter's, which would become Christianity's foremost church, on the spot where tradition has it that Peter was crucified. Furthermore, Christian festivals began to replace old pagan festivals in the Roman calendar. The most famous of these was Christmas, which was introduced to replace the pagan Roman mid-winter festival of Saturnalia and Sol Invictus on 21 and 25 December (for the return of the Sun after the winter solstice) around AD 336. And in the sixth century, the Roman calendar itself was changed. Instead of reckoning its years *ab urbe condita* – from the legendary date for the founding of Rome by Romulus and Remus – it switched to dating events from the birth of Christ. The scholar who carried out the work, Dionysius Exiguus (Little Dennis), calculated that Christ had been born 753 years after the founding of Rome, although we now know he was in error by four or five years. Even so, all subsequent dates ran in the Latin *anno Domini* (AD), or 'after the Year of the Lord', and we still use Dionysius's reckoning today.

By the sixth century, in fact, Christianity had become the dominant spiritual presence throughout the old Roman world, and the Church took on an organisational structure that absorbed important components of provincial Greek and Roman administration. The *episcopi*, or Christian bishops, for instance, were originally Greek and Roman administrative overseers or inspectors; and the *dioecesis* – a managerial unit or 'housekeeping' district of a Roman governor – became the *diocese*, or jurisdictional area for Christian bishops. Even the liturgical vestments of Christian priests with their long robes borrowed components from the official dress of Roman and Byzantine civil servants. And when the Emperor Constantine moved the official heart of the Empire from Rome eastwards to the old Greek city of Byzantium (but now reborn as the fabulous *Constantinopolis*, or 'Constantine's City') in AD 330, the Christian faith also took official root. In Constantinople, however, Christianity came to develop different traditions from those of Rome, and while remaining the same faith, with the same Bible, it nonetheless disputed the jurisdiction of the Pope as Christ's successor on Earth. This eventually gave rise to the Greek Orthodox tradition within the Christian community, in contrast with the Roman Catholic tradition, which was to dominate Latin Europe down to Martin Luther's Reformation in 1517–20.

By the early medieval period the Christian tradition of heartland Europe was Roman Catholic, in Byzantine Greece it was emerging as Orthodox, and in other Eastern regions such as Egypt and Persia it was Coptic or Nestorian. Christianity had effectively taken over the education system, and its priesthood – especially in its upper echelons, such as bishops – embraced some of the finest minds of the age. And while the rise of Islam after AD 622 temporarily destabilised this world, and robbed it of important territories such as Egypt and Spain, the monasteries and cathedrals of Christian Europe were to become major seats of learning, and self-conscious inheritors both of Christ's Church on the one hand and the culture of the world of the Caesars on the other. And that came to include, amongst other things, the scientific knowledge of antiquity.

But had there been any especially *Roman* contributions to scientific ideas? While the Romans as a people had lacked that fascination with pure intellectual inquiry which had made the Greeks so remarkable (Roman intellectual inclinations had been more concerned with the arts of statecraft, law, poetry, history, architecture, town planning and civil engineering), they absorbed what they found useful from the Greeks, and also produced several enduring scientific thinkers of their own. And this began during the pagan, pre-Christian period of Roman antiquity.

The *De Medicina* (*On Medicine*) of Aulus Cornelius Celsus (*c*.AD 30) was destined to become one of the most influential treatises on clinical medicine ever written, and the first to be written in Latin, although its roots were firmly Greek. Even so, it is still fascinating to read today – not least for the ingenious common sense and lack of superstition which runs through it. Also in medicine and anatomy were the works of Claudius Galen (*c*.AD 129–199), a young Greek from Pergamum. Galen had obtained a first-class practical training when he became doctor to the gladiators in the local arena, before going on to Rome and becoming court physician during the reign of Marcus Aurelius. He dealt with every kind of injury that could be inflicted upon superbly fit bodies. Galen laid the foundations of scientific anatomy, and his Latin-titled *De Usu Partium* (*On the Use of Parts*) and other treatises, after their translation into Latin after about 1280, came to form the cornerstone of the medical courses of Europe's Renaissance universities until around 1600.

Outside medicine, there were several Romans who wrote major treatises on the natural world, and one the most ingenious of these was Titus Carus Lucretius (*c*.99–*c*.55 BC), whose *De Rerum Natura* (*On the Nature of Things*) combined an awareness of Greek achievement with his own independent ideas. Lucretius was the great Roman populariser of the largely forgotten atomic theory of matter put forward by the Greeks Leucippus and Democritus four centuries earlier. He argued, against Aristotle, that things were not made up of four elements, but of tiny homogeneous particles. The way in which these particles agglutinated and combined together formed every substance that could exist. And as an Epicurean philosopher, he argued that the wise man or woman would cheerfully accept mortality, and realise that death was no more than a natural dissolution of the body into its component atoms, involving no pain or suffering, for Pluto's gloomy and terrifying kingdom was a fiction. Fear of death, indeed, was the source of much misery on Earth.

Lucretius was therefore one of the first sustained exponents of scientific materialism, in which the world was held to consist entirely of eternal atomic building blocks, the chance combination of which produced both ourselves and all things that exist. No *logos*, or *nous*, or Creator God was really necessary – only time and chance.

The most famous and most widely read of the non-medical Roman scientific writers was Gaius Plinius Secundus, or Pliny (AD 23–79), known to history as Pliny the Elder. He was a Roman aristocrat, a soldier, an admiral in the Roman navy, a humane gentleman, a collector of books, and a shrewd observer of nature. But what assured his historical memory was the mode of his death, which took place when he was overcome by noxious fumes as he was trying to organise the evacuation of the stricken Italian cities of Pompeii and Herculaneum during the great eruption of Vesuvius in AD 79. His observations of the earlier stages of the Vesuvian eruption, and especially of the formation of its colossal plume of dust and ash – later committed to written record by his nephew, Pliny the Younger, and by others who survived – has also given him a special new relevance today. In their attempts to understand what happens during a massively destructive volcanic eruption, modern vulcanologists have found Pliny's observations on the sequence of the AD 79 eruption to be invaluable.

But Pliny's great scientific achievement was his *Naturalis Historia* (*Natural History*) of AD 77, an encyclopaedic treatise on many aspects of the natural world. This Latin work was to become the leading European source book right through the middle ages, explaining as it did a wide range of phenomena, from the nature of wind and weather to the behaviour of animals and the characteristics of plants and minerals. Pliny's *Naturalis Historia*, moreover, was to form the point of departure for a whole genre of medieval quasi-scientific literature, such as the *Bestiaries* and *Lapidaries*, which discussed respectively the nature of animals and minerals, but now in the light of Christian-related stories. (The eagle, for instance, was said to be the

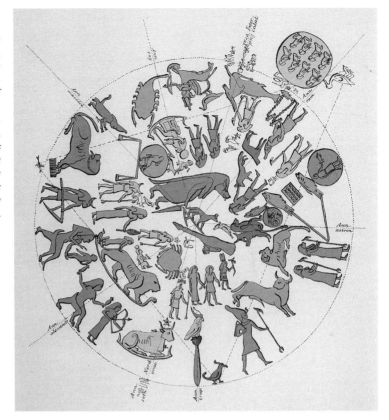

The Dendera Zodiac, from the ceiling in the chapel in the temple at Dendera, Upper Egypt. The zodiacal stars form the inner groups of constellations depicted as stylised human and animal figures. The Hippopotamus included part of the Plough. The thirty-six figures around the edge represent the sky 'decans' or divisions.

Part of the inner group of trilithons at Stonehenge. Although Stonehenge is characterised by some astronomical alignments, such as the midsummer sunrise, we do not know the original purpose of the monument.

An armillary sphere, European, 1605. The inner sphere represents the Earth, and the broad brass band the zodiac. The sphere could be used to solve problems in spherical trigonometry.

An Islamic astrolabe, 1605. The astrolabe was a 'flattened' and more portable version of the armillary sphere, used for solving problems in celestial geometry.

Jai Singh's observatory, Delhi.

The Samrāt Yantra (above) *is an
enormous sundial. The inclined
staircase runs north-south and points
to the north pole. It is 68 feet high and,
depending on the time of day, casts its
shadow on to one of the two east-west-
lying stone quadrants, each of 49½ feet
radius. The instrument could measure
the Sun's position in the sky with great
accuracy, as well as tell the time.
The Ram Yantra* (right) *is a gigantic
masonry cylinder 54 feet across and
24 feet high. It is divided into thirty
radiating divisions of 6 degrees each,
and was used to measure the exact
horizontal and vertical angles of
astronomical bodies.*

Hipparchus (130 BC) as envisaged in the over-charged romantic imagination of Alexandre de Bar. The armillary sphere, as depicted, could never have worked in practice, while Hipparchus is shown using a telescope of Victorian dimensions.

St. Augustine in his cell, by Sandro Botticelli (Ognissanti, Florence, 1480). The painting draws together perfectly the images of science and religion, for the Saint has his bishop's mitre, along with a brass armillary sphere, a mechanical clock, and a copy of what is probably Euclid.

The medieval microcosm. Salisbury Cathedral's mechanical clock (1386), whereby machinery replicated the Sun's motion around the Earth.

The nave of Salisbury Cathedral, c.1220. One of the finest examples of early English Gothic, with its soaring height, vast interior space and great windows, made possible by the load-bearing properties of pointed arches and flying buttresses.

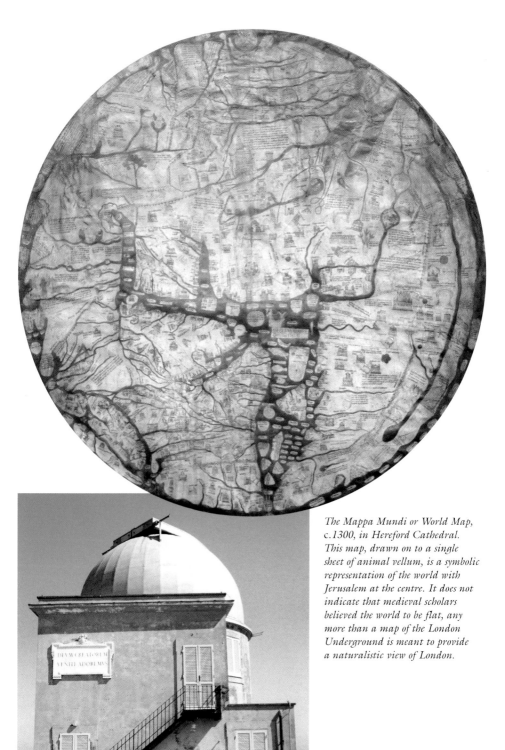

The Mappa Mundi or World Map, c.1300, in Hereford Cathedral. This map, drawn on to a single sheet of animal vellum, is a symbolic representation of the world with Jerusalem at the centre. It does not indicate that medieval scholars believed the world to be flat, any more than a map of the London Underground is meant to provide a naturalistic view of London.

The Vatican Observatory, near Rome. The observatory has been in its present location, above the Pope's summer residence at Castelgandolfo, since being moved out of Rome in 1933. The dome houses a twentieth-century astrographic camera by Zeiss.

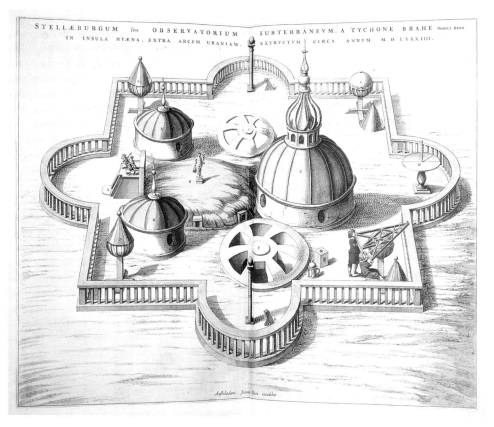

Part of Tycho Brahe's observatory at Hven, Denmark, c.1590. Though Tycho was working before the invention of the telescope, the various domes housed large brass instruments used for measuring the angles between stars. (Jean-Loup Charmet, after an engraving in Atlas Blaeu, *1665.)*

A nineteenth-century illustration of the pyramids of Giza, Egypt.

The allegorical frontispiece of G.B. Riccioli's Almagestum Novum *(Bologna, 1651). The goddess Urania holds a balance, in which the Earth-centred cosmology of Tycho Brahe is found to be more weighty than the Sun-centred Universe of Copernicus. The figure on the left holds a telescope, while the hand of God points down from above.*

noblest of birds, for it was believed not only to be the highest flyer, but also the only one capable of looking at the Sun directly. It enjoyed a special relationship with God, being the only bird to be ranked amongst the four prophetic beasts in the Book of *Revelation* – which probably explains why church lecterns came to be fashioned in the form of a large brass eagle to carry the Church's Bible upon its outstretched wings.)

Roman astronomy was far less intellectually inquisitive than that of the Greeks. There were no Roman observatories that used instruments to build up sequences of data that could be employed to solve theoretical problems, and the Romans seemed content to borrow from the Greeks when they required explanations of celestial phenomena. Indeed, our best sources for Roman astronomy come not from explicit scientific treatises but from poetic works that deal with the nature of the sky, such as Marcus Manilius' *Astronomica* of the early first century AD, and the works of Flavius Magnus Aurelius Cassiodorus (*c*.AD 550). There is also the 'Atlante Farnese' Zodiac of *c*.200 BC – a beautiful marble sculpture depicting the Roman god Atlas carrying a celestial globe upon his back, now preserved in the Museum at Naples. But the constellations depicted on Atlas's globe are the already familiar constellations found in Ptolemy's *Almagest*, including the twelve zodiacal signs and other groups of stars.

Generally speaking, Roman science tended towards the encyclopaedic rather than the inquisitorial: practical knowledge carefully arranged to be of use when solving problems concerning the seasons or finding the time of the night, rather than the devising of abstract models with eccentrics and epicycles to account for the planetary retrograde loops. Indeed, this practical approach to astronomy is best exemplified in that body of writings known as the *Corpus Agrimensorum*, which was a collection of rules and techniques used by Roman *agrimensores*, or land measurers. Here, the emphasis is on the ability of the land measurer to use astronomy to find the time, make sundials, and align buildings, but the work is indifferent to abstruse questions concerning epicycles or the mathematics underlying celestial motions.

But perhaps Roman astronomy's truly long-lasting contribution to civilisation came in 46 BC, when Julius Caesar employed the Greek astronomer Sosogenes to produce an improved calendar. Knowing that the year was 365¼ days long, and not 365 days, Sosogenes inserted a 'leap year' every four years to round up the quarter days. His 'Julian Calendar' was to be subsequently taken over by the Christian Church, and it remained in use in Roman Catholic Europe until 1582, in England until 1752, and in Russia until after the 1917 Revolution, while it still serves as the liturgical calendar for the Eastern Orthodox Church in the present day. But Sosogenes' Julian Calendar was not perfect. He had overestimated the length of the year by a few minutes, and by the time of the middle ages the small annual errors had led to the accumulation of long-term errors. Thus astronomy acquired an increasingly high profile in the medieval Church.

Classical Greek science tended to pass into the late Roman world of the sixth and seventh centuries AD not in the form of primary texts, such as copies of the works of Aristotle or Hipparchus, but in the form of Latin encyclopaedic digests and commentaries. This was probably because, when the Roman Empire was beginning to disintegrate after about AD 400, and Visigothic, Hun, and other barbarian tribes were sweeping into the Italian peninsula, original texts were being destroyed and fresh copies – in the pre-printed-book world – were not being made. Instead, encyclopaedias or digests were compiled, recording the salient features and main arguments of a particular earlier Greek or Roman scholar's work. One of the most celebrated of these was that of Ancius Manlius Torquatus Severinus Boethius (*c.*AD 480–524) – scholar, Roman consul and advisor to the Emperor Theodoric, who was condemned to death for opposing Theodoric's unjust and absolutist conduct.

It had been one of Boethius' ambitions to compile a great Latin translation and digest of the great philosophers of Greek antiquity, although his death sentence made its completion impossible. Even so, his *Consolations of Philosophy*, written while he was awaiting execution, became – like Socrates' *Phaedo*, and Christ's farewell discourses to his disciples before his crucifixion – one of the world's classic statements on the meaing of life and death.

It is not certain whether Boethius was a Christian, though it is believed to have been very likely, for the Church later canonised him as St Severinus, and honoured his tomb in Pavia Cathedral. And there were scholars within the early Church who also undertook the production of digests of classical learning, such as Isidore of Seville (*c*.560–636). Isidore was a Spanish archbishop, whose *Etymologies* was written only a few decades before the Muslims swept across the Straits of Gibraltar and into Spain, and his work was to become a major source for European scholars over the next six or seven centuries.

Greek astronomy, we must not forget, continued to be culti-vated in the Byzantine Empire of Constantinople. For while the capital city of the Byzantine Empire had become Constantine's 'New Rome' in the fourth century AD, it had gradually evolved as a political and intellectual entity in its own right: Orthodox Christian and not Roman Catholic in its religion, Greek and not Latin in its language, and often suspicious of its older cousin, Rome, in the West. In Constantinople, however, many Greek scientific and philosophical books were preserved, and were available for study in their original language, without need for translation. A sophisticated astronomical calendar calculator using brass gearwheels, of *c*.AD 520, has survived, but the Byzantine Greeks were not interested in astronomical *research* in the way that Hipparchus and Ptolemy had pursued it. (A later and less elaborate descendant of the Antikythera Mechanism, described in Chapter 5, is now preserved in the Science Museum, London.)

It has, indeed, long been a subject of interest among scholars as to why the Romans and Byzantine Greeks, who inherited the intellectual mantle of the Greek philosophers, failed to continue their active scientific tradition after AD 150. I believe that several reasons might be suggested. One very important cause was the end of the geographical expansion of classical Graeco-Roman civilisation by the third century AD. And with it, one might argue, the broader urge towards intellectual expansion had reached its limit. With the first barbarian invasions of the Roman world in the fourth century, it was clear that the classical order was, for the first time, being forced on the defensive, and with it came a crisis of confidence in its own values, culminating in the fall of the Roman empire in western Europe in 476. One could also argue that both classical astronomy and medicine had reached their respective peaks of technical development by AD 200. For astronomy in particular to go further, it was necessary for more centuries to pass so that the errors in Greek planetary theory could accumulate and become sufficiently large for the Arabs to measure with their new, large instruments by AD 1200.

Yet one further major factor must have been the nature of Christianity itself. When Christ ascended into heaven, forty days after his resurrection, his disciples had understood that he would return to Earth soon after, to bring about the Last Judgment consigning the saved and the damned to heaven and hell respectively, and then, after a time, to destroy the physical world, and return it to the nothingness from which God had originally made it. And even when this Last Judgement and destruction failed to take place century after century (for Christ had never given a date or time), the physical world still seemed an ephemeral place of uncertain future duration. For these were, indeed, what most Christian scholars believed to be the 'last days' of the world; and as such, it was far more worthy to repent and seek salvation than it was to investigate the geometrical anomalies of the lunar orbit.

If these really were the 'last days', moreover, should one spend them reading Plato and Aristotle who, men of profound insight though they might have been, were still pagans? The status of pagan Greek and Roman knowledge was to trouble many Christian intellectuals of this time, especially men like Origen (AD 185–254) and St Augustine (AD 354–430), who felt torn between their faith and their love of the pagan classical writers. The most influential of these scholars was St Augustine, whose writings were to have an enormous effect upon the intellectual development of Christianity.

Augustine was born at Tagaste in north Africa, into a well-to-do family. His father was probably a pagan, though his mother Monica was a Christian. Being a natural scholar, he took to books from an early age and fell especially in love with Cicero's glorious Latinity, and with Plato's ideas, although he also loved women and kept a mistress. And then, this leisured young scholar about town had a crisis of conscience. Being in Italy, he was baptised a Christian by Ambrose, Bishop of Milan in AD 387, and began upon a wholly new direction in life. Augustine was the first, and probably is still the greatest, philosopher that the Christian Church has ever had. He was to wrestle with problems of scholarly authority, and with the true spiritual value of his beloved Plato, Cicero and other writers who had, in themselves, lived exemplary moral lives, and who had seemed to prefigure Christianity's concern with eternal and transcendent truths up to 400 years before Christ had been born. For had not Socrates, Plato, Cicero and others taught men to look beyond the world of fleeting shadows to a realm of perfect Forms that lay in the mind of the *nous*, and which could now be interpreted as an honest pagan's glimpse of Christ's heaven? When the Visigothic chieftain Alaric entered Italy and sacked the eternal city of Rome in AD 410, Augustine saw this not only as a major turning point in history, but as a divine judgement upon worldly power in these last days. It inspired him to write, between 413 and 426, one of the greatest books of late classical antiquity: *De Civitate Dei* (*The City of God*). (Alaric, interestingly enough, was not a pagan, but, like many Goths, was an 'Arian', or subscriber to the belief that Christ was not an eternal aspect of God, as orthodox Christians see him manifested within the Trinity, but a special man of supreme righteousness.)

We also know much about Augustine's life and the inner workings of his mind from his autobiographical *Confessions*. But the *Confessions* too display Augustine's stature as a philosopher, for in chapter 11 of that work one finds a discussion of the nature of time which is remarkably relativistic and almost seems to prefigure Einstein.

Time fascinated many early Christian intellectuals, and Origen and others had written about it before Augustine. Was time the same in all places? And did the souls of the dead, or the angels, have a similar sense of time to that of mortals on Earth? It seems that certain pagan philosophers had been testing their Christian colleagues by asking: 'Why did not God make the world *sooner* than he did?' Why, indeed, did an eternal God wait for so long before starting the process of creation, if the world was only four or five thousand years old, as it was then believed to be?

What Augustine argued, however, is that God himself is timeless, and lives outside time. Time, indeed, can only exist after God has created material things, such as astronomical bodies, and set them moving in space. Therefore, as time could not have existed before God made the heavens, He could not have made the Universe sooner, because no prior time frame existed.

Quite apart from philosophical discussions about time, however, the Church has always needed the services of astronomers to determine the annual date of the supreme event in the Christian calendar: Easter Sunday, which is celebrated to mark Christ's resurrection. But Easter Sunday is a moveable feast, and can occur between 21 March and 25 April, due to an astronomical mechanism governed by the falling of the full Moon after the spring equinox. It therefore has obvious parallels to the astronomical mechanism from which the moveable Jewish festival of Passover is calculated, for Christ's actual crucifixion took place, according to the Gospels, on the Friday before Passover Saturday, so that His resurrection took place on the Sunday following it.

In the early Church, Easter was sometimes celebrated at the same time as the Jewish Passover, but this synchronicity later changed, and several Church Councils and scholars were active in developing a reliable computational formula to arrive at the date of Easter each year. In consequence, a set of elaborate rules, based on the 19-year Sun–Moon cycle and other factors, were gradually perfected, as it was deemed essential for all Christians everywhere to celebrate Easter on the same day, at least within their respective Catholic or Orthodox Churches.

But this was a formidable task, for as the ancient Greeks had known only too well, the Earth–Moon–Sun relationship, with all of its epicycles and eccentrics, was immensely complicated. Using tables adapted from Ptolemy and other astronomers, however, it became possible, by the seventh and eighth centuries AD, to calculate Easter dates for cycles of years into the future, and thereby to establish precise liturgical calendars. Much of the initiative in this respect came from England. The Synod of Whitby, Yorkshire, of AD 664, in which prominent churchmen had met together to settle important ecclesiastical matters, led to the regularisation of calendrical reckoning for Easter, and to the introduction of the *anno Domini* dating system, at least in Britain. We know this from the writings of the Venerable Bede, who subsequently wrote an historical account of the Synod while living memory and possibly written records of it were still available.

Britain's first astronomer of international standing was a monk of Jarrow. The Venerable Bede (AD 675–735) never seems to have travelled beyond the locality of his native Northumbria, although his friend the Abbot of Jarrow, Benedict Biscop, made several journeys across Europe, and never failed to have copies of interesting books that he encountered sent to Bede and the Abbey library. Bede wrote a *History of the English Church*, a life of his fellow Northumbrian, St Cuthbert, and several works relating to astronomy. As an educated man, Bede of course knew the classical arguments for a spherical Earth, which he envisaged, as had Ptolemy, as standing motionless in the centre of the Universe with the spheres of heaven rotating around it. Living in a monastery that overlooked the North Sea, he was familiar with the relationship between the Sun, Moon and tides. Considering the ethos of religious brotherhood that underpinned the very life of a monastery, Bede would not have kept his learning to himself. It would have been passed on to his brothers, taught to the young monks and novices, and written down for the use of priests and monks elsewhere, and for posterity, and Bede's *De Arte Metrica* (*On the Art of Measurement*) was almost certainly written as a teaching book.

Bede's manuscript *De Temporibus* (*On the Times*) provided rules for performing various types of calculations that were of particular use in fixing liturgical dates, and his work on finding the date of Easter within a cycle of years came to be employed across Europe. Yet we know nothing about Bede's instruments, or even whether he had any. On the other hand, while most of his work was based on tabular analysis and calculation that used earlier observations, it is difficult to believe that Bede did not have a method of checking his figures against the actual movements of the Sun and Moon amongst the stars.

By the time that Bede was working in the early ninth century, moreover, one particular factor would have become especially noticeable when it came to the production of accurate calendars: the precession of the equinoxes (discussed in Chapter 5). From the early days of the Church until Bede's time, some seven centuries had passed by, producing an inevitable conflict between the civil calendar and the sky itself. For instance, while the equinoxes and solstices were originally found by Sosogenes in 46 BC to fall on the 25th day of March and September, June and December respectively (with occasional adjustments for leap years), precession meant that they kept falling slightly yet inexorably before these dates; only by a tiny fraction each year, it is true, and undetectable over short periods, but after four centuries this precessional error amounted to just over three days. Therefore, in AD 325 the Council of Nicaea reset the spring equinox date, crucial for calculating Easter, at 21 March. But by the time of the Synod of Whitby, and then of Bede, the equinox was falling yet another three days before the calendrical date.

This precessional backsliding was fundamental to fixing the date of Easter in a given cycle of years. There was no problem, of course, if the spring full Moon, or Paschal Moon, fell on Sunday 21 March. But what if the Paschal Moon fell on Sunday 19 March when you knew, from the rate of precession, that the true *astronomical* spring equinox fell on 18 March? Did you calculate Easter slightly before the calendar date that year, or did you instead let a whole lunar month go by, and calculate the feast in late April?

While one might argue that the date itself was not especially important – what really mattered was that Easter should be celebrated by all Christians at the same time – we should not forget that in medieval Europe, communications were relatively slow. Astronomers in Rome, Constantinople, or some other great religious centre, could not perform an official calculation and ensure that every church in Christendom – including those in outlying areas, such as the Western Isles of Scotland – would receive the message in time. Instead, astronomer priests attached to cathedrals and abbeys across Europe would carry out their own calculations based on accepted formulae and tables. Astronomers, therefore, fulfilled a crucial liturgical function, and it was partly for this reason that Bede's work became so influential.

It is ironic that during this period, when the Roman empire was disintegrating in southern Europe under a combination of economic pressures and barbarian invasions, the monasteries of the north were enjoying a time of cultural and spiritual flowering. Mountains and islands were important to this renascence, no doubt, for the resulting isolation gave them peace and security. There was, of course, Bede and his colleagues in Jarrow, the Holy Island of Lindisfarne, and the other Northumbrian monasteries. In Switzerland, the great monastery of St Gallen, traditionally said to have been founded by the seventh-century Irish monk Gallus, became a byword for scholarship, while Ireland herself first won at this time a reputation for being a 'land of saints and scholars', her most spectacular surviving artefact being that set of illuminated Gospels known as the *Book of Kells*, now preserved in Dublin. But alas, this renascence was cut short in those places easily accessible to the sea, following the sudden appearance of Danish and Viking raiders in the late ninth century. Jarrow was looted in 867–70, and Lindisfarne in 875, while the craggy ruins perched atop seemingly inaccessible cliffs around the coastline of western Ireland are all that remain of the communities of these first pre-Viking Irish scholars.

And yet, as the Roman empire ceased to exist, a new energy began to manifest itself in northern Europe. Not only had Christianity taken a firm root in the old barbarian lands north of the Alps, but it now became their principal engine of civilisation. New regional and political identities were also forming: France, Germany, Austria, and the British Isles with their own tribal districts of Scots, Welsh, Irish and English, each with their own rulers, bishops, abbots, and priests, invariably became not only the spiritual guides of these fledgling states, from the Bay of Biscay to the Russian steppes, but also civil servants, keepers of records and calculators of calendars. And these men, still wearing professional garments based upon those of Roman officials, and using the Latin tongue of the Caesars for their daily business, were deeply aware of their origins: a once-pagan empire that had ruled the known world, which had first crucified and then been overtaken by the spirit of Christ, whose second coming they all awaited. In the meantime, good order and civilisation were to be maintained in the world, learning encouraged, and the forces of darkness, chaos, and primitive superstition driven away.

Then, on Christmas Day, AD 800, the Frankish king Charlemagne, or Charles the Great, was crowned Emperor of the West in the Cathedral at Aachen (Aix-la-Chapelle), Germany. He was Caesar revived, the ruler of an empire that would run from the Pyrenees to the Danube, northern Italy and the Black Sea (though excluding the British Isles), whose aspirations would become the genesis of so much mainland European history. Charlemagne's Holy Roman Empire saw itself as taking the best from the ancient Caesars, and combining it with a thoroughly Christian basis for society. Knights would wage war with the heathen, protect the Church, and, as it was understood in that feudal hierarchical society, administer justice; while monks and priests would pray and promote Godly learning. Charlemagne and his successors would wield the *imperium*, the just authority of the Caesars, whereas the Popes in Rome would guard the keys to the gates of heaven which tradition said Christ had entrusted to St Peter and his successors. This spectacular vision of a Christian state would undergo many ups, downs and dilutions, and would last for exactly 1,006 years, until Napoleon Bonaparte and his French Revolutionary storm-troopers abolished what survived in 1806.

Charlemagne – like his later near-contemporary, Alfred of England (849–899) – recognised the importance of learning to the administration of a well-run society, and he began to attract both scholars and books to his Frankish court long before he was crowned Emperor at the age of 58. His real coup was an Englishman, Alcuin of York (735–804), who became Charlemagne's Master of the Schools.

Alcuin, who was a Yorkshireman by birth, had, as a young novice, entered the already famous school attached to York Minster, and received his education under Ethelbert, whom he succeeded as head of the School in 767. At this time, however, even the library of a great cathedral church would have contained only a small range of book. Beyond copies of the Bible, St Augustine, and other Church Fathers, there would have been editions of late Roman books on Latin grammar – such as those of Donatus and Priscian – some of Boethius's writings, and 'encyclopaedic' fragments on classical medicine, astronomy, animals, plants and minerals, together with Bede's texts on calendars. A library would be regarded as major if it boasted forty individual titles. We know that York Minster possessed copies (some fragmentary) of Aristotle, Pliny, Cicero and Virgil, as well as religious texts, while Alcuin was fortunate enough to personally own at least fourteen books.

In this world of ninth- and tenth-century northern Europe, therefore, cathedral schools and monasteries, along with the itinerant Imperial Court, had become the main foci of learning. It was first and foremost, however, a literary culture based on preserving what fragments remained from ancient Greece and Rome, combined with the new literature of the Christian Church. Its intention was what we might deem conservative, rather than investigative, insofar as it aspired to conserve and pass on intact that which was left from the past without losing it, while keeping men's and women's minds fixed upon preparing themselves for a life beyond the present.

At the heart of this new culture was a common language of learning: Latin. The language of the Caesars had also become – after those inevitable changes which will take place within any living language – the international *lingua franca* of the Church, the law, public administration, diplomacy, and the academic world. A Danish bishop could write to an Italian counterpart without the need for a translator, and a student from Scotland could attend lectures in a French cathedral school and not encounter any language barrier beyond that of differing regional variations in the pronunciation of Latin. This situation, moreover, would continue down to the sixteenth and seventeenth centuries, when Europe's vernacular languages finally rose from the streets and into the pulpits and lecture theatres, and began to erode some of that common intellectual and spiritual culture which had been one of the glories of Christian medieval Europe.

But how did the north European Christian world relate to the flowering of Islam? On one level, it was perceived as a threat, for Islam, like Christianity, was not an ethnic religion but an evangelical one based upon the acceptance of a belief system and a submission to God that was open to all men and women, rather than one which was the inheritance of a particular race of people. Indeed, Islam's sweeping up of Egypt, north Africa, Palestine and Spain had probably terrified the West, which not infrequently saw evangelical Islam as a scourge sent by God to punish sinners and herald the end of the world. By the tenth century, however, matters had begun to settle down somewhat, and certain Europeans even took an interest in aspects of Islamic culture; for while the two religions differed sharply on the subject of the divinity of Jesus Christ, they nonetheless acknowledged the supremacy of the one Creator God, not to mention their mutual historical links with the Jewish patriarchs.

Christian Europe's connections with Islam therefore came to develop in two very different ways: one intellectual, and the other military. In places like Sicily, southern Italy, Majorca, and other Mediterranean islands, Christian, Jewish and Muslim groups learned how to tolerate each other, and to trade and exchange ideas. It was probably through these contacts that it came to be realised that the Muslims had Arabic translations of whole works of ancient Greek authors who were known only by name or in fragments in Paris or St Gallen. As a result, beginning with a tiny trickle and growing to a flood, complete classical authors began to pour into the West, and to be translated into Latin (as we shall see in Chapter 8).

It is very probable (although there is no conclusive proof) that before this tide of translation got under way, a young French priest, who would end his days as Pope, actually visited Muslim Spain, obtained scientific manuscripts, brought an Arabic astrolabe into France, and built a water clock. His name was Gerbert of Aurillac (*c*.940–1003). And even if Gerbert did not visit the Muslim regions of Spain in person, he was certainly familiar with its astronomical achievements, having received part of his education in the Christian college at Vich, near Barcelona, where his talent for mathematics won him attention. Gerbert later became Archbishop of Ravenna, Italy, and on the death of Pope Gregory V in AD 999, the German Emperor Otto III was instrumental in his election as Pope. For his papal title Gerbert took the name Silvester II, and he reigned to his death in 1003.

It was not, indeed, without significance for the future history of science and religion that the man who headed Christendom at the dawn of the second millennium was not only a learned priest, but an astronomer and teacher who had glimpsed the scientific treasures that could become available to western Europeans through contacts with Islam.

However, in addition to purely academic contacts, Europe's dealings with the Muslim world after about AD 1000 were becoming increasingly war-like. By this time, indeed, the explosive evangelical dynamism of early Islam had largely expended itself, having been stopped at the French Pyrenees in the west, and by the great Christian 'buffer zone' of Byzantium in the east. But by 1000, western Europe was preparing to strike back with the beginning of the Crusading movement.

This movement started in the far west of Europe with the 'Re-Conquista' – the taking back of those regions of once-Christian Spain that were now ruled by Muslims – and was triggered in part by the increasing Muslim persecution of resident Spanish Christians.

Aristocratic second sons from north European military families realised not only that land was to be had for the taking in Spain, and new princedoms to be carved out for themselves, but that their 'Re-Conquista' was also spreading Christian influence. Indeed, Spain would become a battleground between invading north European Christians and now settled Islamic kingdoms for several centuries to come. This was the time of Rodrigo Diaz (1043–1099), the Spanish folk-hero better known as El Cid, whose legendary exploits would be reworked to produce the material for a Hollywood blockbuster in the 1950s. But the real 'Re-Conquista' carried on until Ferdinand and Isabella of Spain overcame the last Muslim stronghold in the Iberian peninsula at Granada in 1492.

Then, in 1096 Pope Urban II preached a sermon at Clermont in France, exhorting the European knights to go to Palestine, overthrow the Muslim rulers, and seize Jerusalem and all the holy places associated with the life of Christ. And by way of instant response, not only did knights and soldiers from almost every country in Europe heed Urban's summons, but there were also spontaneous crusades of peasants, children and other ordinary folk. Though the colossal loss of life among the crusaders cannot be given in exact figures – only in tens of thousands – Jerusalem eventually fell to the crusading knights on 15 July 1099. After a monumental massacre of the defenders, western Europeans were to rule Jerusalem as a Christian kingdom for 88 years, until that courtly and debonair Egyptian Muslim Prince, Sal-a-din, won it back for Islam in 1187.

In spite of the often sanguinary nature of the contacts, however, East, ironically enough, was meeting West, and Christian was meeting Muslim. And once the embattled French and German knights had seized their chunks of Spain, or won territories in the new Christian kingdom of 'Ultramar' in Palestine, they started to learn from the locals. They learned, amongst other things, that in a hot climate, cotton clothes were more comfortable than thick layers of unwashed wool, and that bathing could be pleasurable. They also found new fruits, such as oranges, discovered sugar to be far more abundant than in Europe, were delighted by the sounds of Arabic musical instruments such as the lute and rebeck, and realised that Muslim architects had discovered that the load-bearing properties of the pointed arch were superior to those of the Roman circular arch – a feature which they would exploit and incorporate into the soaring new Gothic cathedrals that were beginning to spring up across northern Europe after 1200.

But as far as the history of science was concerned, they discovered books: complete works of the Greek writers on astronomy, mathematics, optics and the medical sciences, all in Arabic editions, together with Arabic works. These were the 'old books for new minds' which, when translated into Latin, would in less than a century change beyond recognition the sum total of knowledge available to Western scholars, and usher in Europe's 'twelfth-century Renaissance'. For now, for the first time, complete texts from the finest minds of pagan antiquity could be studied from the Christian perspective of the one Creator God; and while the men of Paris and Oxford did not necessarily like Islam, they were nonetheless willing to admit that Muslim scholars could be clever and sincerely religious, and that it was worth listening to what they had to say. And lying at the heart of this Renaissance was the eternal beauty of the heavens: how something of such ineffable beauty could contain within it such mathematical complexity, and how the sinful men and women of this world could unravel that complexity using the divine gift of reason, and finally rise up through the celestial spheres to enter the glorious City of God.

EIGHT

With angels ascending the spheres:
the medieval cosmos

One of the great questions which faces historians is why cultures
suddenly rise, seemingly from nothing, to an era of brilliance that
changes the course of civilisation, only to pass into decline. The
Greeks after the sixth century BC, and the Arabs after the seventh
century AD, are prime examples of these 'renaissances'; and while one
can identify all sorts of causal factors – such as economic, political and
topographical – one can never quite explain the unique concatenation
of circumstances that leads to one culture simply prospering, and
another suddenly exploding with new ideas and ways of doing things
that change the future course of the world.

Europe in the twelfth century was one such culture. From an
impartial perspective in, let us say, 950, one could identify great
strengths in Europe: the Christian faith with its spiritual confidence,
increasingly stable monarchical governments based upon Roman and
feudal patterns, and a large number of cathedrals and abbeys with
groups of young men studying Latin grammar, theology, and fragments of classical science. But in many ways the Europe of 1250 –

with its crusading expansiveness, its new Gothic architecture, its burgeoning mercantile life, and the prodigious intellectual energy of its new universities – had become a place altogether different from that which it had been two or three centuries earlier.

One major factor in this remarkable ascendancy of European civilisation was growing political weaknesses in the Muslim world. Internecine divisions between the ruling groups of the Near East, not to mention something of a backlash by the Christian communities of Spain against the increasingly harsh policies of their Muslim overlords, had exposed weak spots which the crusading armies of the eleventh century could successfully exploit to advantage. Central to the rise of Europe had been the new politics of feudalism, in which powerful bonds of personal and dynastic loyalty had been built around a new style of warfare, paid for by a new form of land tenure: the feudal *feoffdom*, or land grant given to men who would form the officer class of the medieval armies. These men were the knights.

While a wealth of romantic mythology would subsequently grow up around knighthood and chivalry, feudal knighthood was, at heart, about a new style of warfare: not conducted on foot in tightly disciplined legions of infantry, as had been the key to Rome's success, but rapidly moving, on horseback, and often engaged in freestyle by individual knights. The landed feoffdoms, complete with the peasants who would generate the agricultural surplus to support the system, provided the knight with the freedom necessary to perfect his murderous fighting skills, and to breed the biggest and most powerful horses that the world had ever seen. The key to this new horse warfare was the invention of the foot stirrup – a device not known to the Roman empire. A stirrup provides a rider with more stability and greater freedom of movement in the saddle. He can charge an opponent without being himself unhorsed, stand up in his stirrups, take fearful swipes with a four-foot broadsword without falling off, and wear flexible armour that is strong enough to protect him against most weapons.

Though the Muslim armies of Palestine and Spain used horses, they were smaller and less powerful animals than those of the north. These were fast Arab racehorses, not the slower-moving yet indomitable massive beasts that carried the Frankish knights. Their Arab riders wore little or no armour, and depended on rapid descents upon their foes where their arrows, javelins and scimitars would wreak sudden, paralysing havoc. But when such men were charged in turn by a regiment of carthorses, with each horse carrying a metal-clad man with a 15-foot lance, the sheer impact was profoundly destabilising to one's strategy, to say nothing of the contrasting effects of razor sharp scimitars striking steel plate and battle-axes striking cotton-clad bodies.

It was this new style of fighting, and the feudal economy that supported it, which was to give Europe a new confidence and expansive potential. On the one hand it gave rise to attempts to impose Christianity by the sword, while on the other it created an economic system of great flexibility. Feudalism could not only support soldiers; it could also be adapted to endow and maintain churches, universities and other scholarly enterprises. As Bishop William of Wykeham was by no means the first to show, when he founded Winchester School and New College, Oxford, in the 1380s, the profits of feudal land tenure could support the studies of 140 'poor clerks' a year between his two foundations, with seventy at each, and provide each one with a free education.

Indeed, we often forget that the cathedral schools and then the universities of medieval Europe were invariably *free* places of education. Generally speaking, their students were the bright sons of farmers or small urban craftsmen, whose talents had been spotted by the local priest or schoolmaster. A lucky lad would obtain a free place at Paris, Bologna or Oxford, depending on where he lived. He would go on to perfect his Latin, read the pagan classics of Rome and Greece, along with the Bible and Christian Church Fathers, and after graduating, embark upon a career that led him a million miles away from pigsties or fishmongering. He might become a university don, a judge, a senior civil servant, a powerful bishop, or – like William of Wykeham himself – all four!

Yet why should a man who had risen from the farmyard to be an advisor to the King wish to spend his money on educating the rising generation or founding almshouses for paupers? I would argue that the Catholic doctrine of Purgatory had much to do with it. Purgatory was a place to which the souls of the dead descended to await final Judgment and be consigned to heaven or hell. Depending upon how one had lived one's earthly life, Purgatory could vary between the uncomfortable and the downright agonising, as was so brilliantly described by the Florentine poet Dante Aleghieri in his *Inferno* and *Purgatorio* (1314–21).

If one had risen from obscurity to wealth and power, one was likely to have been driven by the sins of pride, greed, avarice, lust, and so on, and to have trampled on lesser men and women, and therefore be likely after death to face a very painful Purgatory and perhaps eventual consignment to hell's everlasting fires. How, therefore, could the rich and famous avoid this fate? They could do so by performing what were called 'works', or acts of charity, such as founding schools, colleges, and hospitals, whereby they used their ill-gotten gains to benefit those poor people to whom Christ himself had ministered. But not only were the basic resources of cash and land provided. The charters that created charitable institutions across the Christian world often specified a requirement from their future beneficiaries that they remember and pray for the soul of their long-departed benefactor. And as God heard their daily prayers – often said in chapels especially built for the purpose, with the sacraments administered by a salaried priest – he would look kindly upon the soul of the old, successful sinner, ease his purgatorial suffering, and hopefully grant him the final reward of a place in heaven. The soul of

William of Wykeham, for instance, has already enjoyed the benefits of six centuries of directed prayers said on his behalf by Wykeham scholars in Winchester and Oxford.

At the heart of this process was the radically egalitarian view of the nature of the human soul in Christianity. Unlike the religions of the Greeks and Romans, in which the successful could be semi-deified and rise above common mortals, worldly success in Christianity always brought with it the taint of sin and the possibility of damnation. After death, the king and the beggar, the bishop and the outcast leper woman stood on equal terms before God at the Last Judgment. Never before had any religion contained a more powerful inner dynamic that encouraged the successful to behave charitably to their less fortunate brothers and sisters, and one of the great fruits of this belief was the foundation of educational charities, whereby future generations of gifted young men could be trained up to serve the Church and its ministry in the world.

One might argue, therefore, that what made medieval Europe so powerful and confident was its ability on the one hand to recognise and mould talent, and on the other to spur it towards the public good via an explicit set of eternal rewards and punishments. For just as some of the greatest intellectuals and scientists of medieval times were scholarship boys who had made good by rising up through a flexible education system, so some of its most daring military men had been the landless sons of hard-up minor gentlemen to whom the murderous freedom of knightly warfare had given an opportunity to 'win their spurs'. As we have already seen, there was plenty of land to grab and even princedoms to establish in the Spanish Re-Conquista or the twelfth-century kingdom of Jerusalem. And all of this military energy, just like the intellectual energy of scholars, was used for the greater glory of God, insofar as it spread the Christian faith into new lands, or else reclaimed old ones, to provide those 'works' whereby earthly savagery might be redeemed to win eternal life.

In spite of this obvious freedom of movement which could be enjoyed by the exceptionally clever, brave and fortunate, European society still stood foursquare upon the naturalness of hierarchy. Hierarchy itself was fixed, even if individual careers were not, and it cannot be denied that the vast generality of medieval people remained more or less in the same social places into which they had been born. Hierarchy was natural, moreover, because it so clearly underpinned God's plan for Creation. Had not even Christ himself enjoined his followers to 'render unto Caesar that which is Caesar's and to God that which is God's', as well as instructing men, women, soldiers and fishermen to be content with their lot in life? Of course, this naturally presupposed that these worldly hierarchies would operate justly: that shepherds (rulers) would protect their sheep (people) rather than turn into ravening wolves and eat them; but any such abuses that sprang from original sin and human greed did not invalidate the divine principle of hierarchy in itself.

Moreover, the concept of hierarchy also lay at the heart of medieval science, for our senses clearly revealed to us the divisions into which God had divided His creation. Furthermore, these divisions were not only physical, but also moral and spiritual, for to the medieval mind, both were inextricably connected.

Creation was seen as consisting of two great realms: the spiritual and the mundane (earthly). In the spiritual realm were God's angels, who were eternal beings with no physical bodies or even ostensible gender, living outside time and continually praising God with instruments and song. These angels, moreover, also acted as God's messengers, and clearly derived from pre-Christian Jewish angelic figures such as Gabriel, Michael, and the Cherubim of the book of *Ezekiel* who had also borne God's messages. However, in the early Church, around AD 500, they came to be arranged into three Celestial Hierarchies by Pseudo-Dionysius the Areopagite (called 'Pseudo' to differentiate him from the Dionyius whom St Paul had converted to

Christianity in Athens, 450 years earlier), with each hierarchy containing within itself three angel types, in a Trinity of Trinities. The angel groups were formalised into Seraphim, Cherubim, Thrones, Dominions, Virtues, Powers, Principalities, Archangels, and Angels, of which only the last two groups had direct contact with human beings. Dionysius's intention – like that of Origen, St Augustine, and many other scholars before him – was to devise a clear reconciliation between classical Platonic philosophy and Christianity.

These nine angel types were also associated with the nine celestial spheres of the medieval cosmos, thereby firmly linking cosmology and religious belief. Beyond these spheres was heaven, where God Himself reigned with the saints and saved human beings, in the spectacular city of the New Jerusalem with its buildings of gemstones, golden streets, and endless glorious light, as described in the New Testament book of *Revelation*.

Below this angelic realm was that of *mundus*, or the world. Its occupants were divided in turn into their own hierarchies of animal, vegetable and mineral, each of which possessed its own 'virtue' or strength. Animals, for instance, were defined by their possession of instinct, and their ability to move and to bring forth young. Yet they were less prolific in their fertility than were plants which, while unable to move or form social packs, could nonetheless overrun and occupy the Earth's surface with lightning rapidity. The 'virtue' of plants, therefore, lay in their abundance. The 'virtue' of stones, in turn, lay in their endurance, for while stone had neither instincts nor ability to reproduce, the rocks of the mountains had preceded the plants and animals in the order of creation, and would remain unchanged to the end of time. Each of these three 'kingdoms', moreover, contained their own inner hierarchies of excellence, for was not the brave lion nobler than the humble yet prolific rabbit, the stately oak finer than mere grass, and the beautiful diamond harder and more brilliant than useful chalk?

Yet none of these things which made up the world possessed the gift of immortality, for their 'souls' were mere defining characteristics – such as 'lionishness' or 'hardness' – rather than conscious spiritual entities capable of having a free-will relationship with their Creator. Such a relationship was unique to man alone, for human beings occupied a special status in the whole of creation. On the one hand, our physical bodies shared movement, sexual reproduction and blind instinct with the beasts; but on the other hand, God had created man, as *Psalm* 8 expresses it, 'a little lower than the Angels, and... crowned him with glory and honour', and our cognisant and immortal souls urged us constantly to struggle with our bestial bodies and strive towards heaven.

Human beings, therefore, occupied a point between the heavenly and the mundane, as the legacy of biology and physics that made us worldly beings was constantly at odds with our immortal, divine inheritance. This creative tension was to lie at the heart of Christian civilisation, and received its most exhaustive exploration in the literature, theology and philosophy of medieval Europe.

Indeed, to attempt to make sense of medieval culture while ignoring this tension, as some writers have tried to do, is to distort history just as grossly as would some future historian in, let us say, AD 3000 who tried to interpret the twentieth century while failing to consider the relationship between politics and economics.

All of these above-mentioned factors – warfare, patronage, expansive energy, and the galvanising power of religious belief – lay behind that new expression of European civilisation which took place after about 1050. Nor should we forget that it was also at this time that Latin translations of Aristotle and the other Greek and Arabic writers were beginning to pour into Europe, to provide puzzles, paradoxes, and cascades of new ideas to challenge the up-and-coming minds of the age, and to reconcile with Christian thought. For as everyone agreed, God could never contradict Himself. And if this was so, the truths of pagan classicism, as expressed in the writings of Plato, Aristotle, and Cicero, *must* spring from the same source that had also inspired St Paul, Origen and St Augustine. It became the task of the medieval academics, therefore, to explore and reconcile these two great intellectual traditions in an enterprise known collectively as 'scholasticism'. And as they were eventually to discover, such an act of reconciliation is far more of an intellectual adventure than simply matching up pagan and Christian texts. It was to occupy the finest minds of the age, and when great minds think, they generate radical new ideas.

One of the most outstanding early figures of this tradition was a young Frenchman named Peter Abelard (1079–1142). As a young scholar, Abelard had succumbed to his passions when he fell in love with his beautiful and brilliant 17-year-old pupil Heloise, thereby giving rise to one of the great romantic dramas of the middle ages. They had a child together, and Heloise's aristocratic brother then wreaked a terrible vengeance by attacking and castrating Abelard. Heloise entered a convent, and Abelard became the most brilliant scholar in the 'Schools' or early university of Paris. But they still kept in contact, and the letters which the increasingly famous academic exchanged with the subsequent Prioress of her convent are amongst the most beautiful specimens of epistolary writing in history, exploring as they do a whole realm of ideas.

Abelard was active in Paris before the impact of Aristotle had enchanted the minds of Europe's scholars, being, indeed, closer in spirit to Plato. And like Plato's teacher Socrates, Abelard saw truth as accessible through argumentative exchanges. Question and answer, or *sic et non*, lay at the heart of Abelard's academic technique; yet the questioning which he advocated was not that of the point-scoring cynic, but of the honest investigator. For Abelard realised that much error, superstition and illogicality was abroad in the world, and that as God was the source only of goodness and of clarity, the errors could not come from Him. To Abelard, therefore, questioning came from a frank desire to know God's intentions and separate them from the ignorant and muddled thinking of mere men. Hence logic, grammar and dialectic became the preliminaries to all true intellectual endeavour, and essential training for all young scholars.

Yet if one thinks of the church, the great cathedral and monastic schools, Latin grammar, argument, the increasingly refined astronomy of Bede's calendar, and Europe's new strengths that derived from feudal institutions, as laying the explosive charge for the twelfth-century renaissance and its consequences, what was the spark that actually set it off and unleashed its energy? I would argue that the spark was the new books that came flooding in, for they placed for the first time a wealth of classical and Arabic science into the hands of European scholars, and, quite simply, presented them with so much more information.

One pivotal event in this respect came as part of the Spanish Re-Conquista, when in 1085 the Christian knights of Alphonso VI took the Muslim stronghold of Toledo, which contained a great library of classical and Arabic scientific and other books. In the decades after 1085, European scholars such as Gerard of Cremona, Hermann of Carinthia, Adelard of Bath, Michael Scot, and many others who had mastered the Arabic language, began to translate these books into Latin. They included Hippocrates' medical works, Aristotle's *Posterior Analytics* (on logic), *Meteorologica* (on optics), and *De Animalibus* (on living creatures), the mathematical works of Archimedes, Euclid's *Geometry*, Apollonius on the geometry of the cone, and the anatomical researches of Galen, along with many other classical scientists. Nor should we forget that Toledo and other Spanish cities also made available translations of numerous Arabic scientists – Arabs, indeed, of the standing of Alhazen, Averroes, Al-Farabi, Rhazes, Al-Kindi, and Jabir (Geber), whose books on astronomy, medicine, natural history and chemistry were coming to be devoured in Paris, Oxford, Bologna, and Europe's other new universities.

Yet Spain was not the only source of new books. The sophisticated courts of southern Italy, where Jews and Muslims often mingled freely with their Christian hosts, also produced Latin translations of old books. Copies of Ptolemy's *Almagest* were translated both in Spain and in Sicily, while his *Optics* became available via Italy. Through southern Italy, Greek editions from Byzantium entered Europe: Aristotle's *Physica* (on physics), Greek manuscripts of his *Meteorologica*, Proclus's *De Motu* (on motion), and so on. And by the thirteenth century, Robert Grosseteste, William of Moerbeke, Bartholomew of Messina and their colleagues were actively translating them from Greek into Latin.

Here was mathematics, medicine, geography, astronomy, political theory, and geometry. But most important of all was the philosophy of Aristotle. For as we have seen, Aristotle was tantalising. On the one hand, his stress on the causal fact that nothing can move without a mover, clearly argued for the existence of a Creator God of some kind. On the other hand, his timeless, cyclical Universe posed problems for Christians (as it had for Muslims), who thought in terms of a divine creation moving through a linear historical time-frame rather than being unknowably remote. There was no indication in Aristotle's writings whether the unmoved mover who had set the Universe, with all its cycles, in motion was still active, or even in existence, for Aristotle did not speak in terms of a theistic being. For this reason, in 1277 sections of Aristotle's writings were banned in Paris's Sorbonne University. Yet Aristotle, quite simply, was too good to lose. Quite apart from the wealth of fresh observational data and factual knowledge to be found in his writings – from how fish breathe to why sunlight is white – was the wonderful connectedness of his thought, which made it possible to logically link one body of knowledge to another. Aristotle's *Politics, Ethics*, psychology (in his *De Anima*, 'On the Soul') and physiology (in *De Generatione Animalium*, 'On the Generation of Animals', a section of *De Animalibus*) were all connected to the doctrine of the mean or balance. A similar concern with things working best in their proper places underpinned his *Physica*, *Meteorologica*, and *De Caelo* (on the heavens), as it did his writings on the four elements and four medical humours, the innate rising of heat and falling of cold, the vital function of the heart, and the exact analytical procedures of logic whereby one could detect intellectual errors and determine the truth. To the logical system-building minds of medieval Europe, Aristotle was a gift, and as they always knew that two truths can never be in conflict, his thought was gradually reinterpreted in order to make any difficult parts acceptable to the Church. St Thomas Aquinas (1225–1274) – the 'Angelic Doctor', and arguably the greatest of all the medieval Schoolmen – laid the foundation for the eventual acceptance of Aristotle. Once that hurdle had been jumped, Aristotle was to dominate the universities of Europe down to the early seventeenth century, to become the pillar of intellectual orthodoxy against whom both Sir Francis Bacon and Galileo Galilei would write so much invective.

Yet what *were* these universities, and how did they differ from the older cathedral schools? For one thing, the universities (dignified with the title *Studium Generale*, due to their willingness to encompass the whole of learning) were born out of that sheer intellectual and broader social ferment which also produced the Crusades, the internecine conflicts between the emperors and the popes, and that new expansive and self-assertive power of twelfth-century Europe. And without a doubt, the wealth of new ideas that flooded in with the translations of new books played a crucial part, as old authorities were questioned and new ingenious arguments were raised. As in Greece in 400 BC, intellectual freedom became an issue for the first time in the Latin West. This involved a freedom to roam across intellectual pastures new, taste the delights of pagan and Islamic thought, and compare the Platonic doctrine of Forms and the truths of Euclid's geometry with the more familiar truths of Christian doctrine.

But where the cathedral schools had been primarily authority-figure-driven, as bishops, abbots, and their staffs had led the young novices by the hand through the familiar Christian writers, the new universities were really student-driven. During the twelfth and thirteenth centuries, the Bishop of Paris, and the Bishop of Lincoln, who officially supervised Oxford, wrestled constantly with tumultuous students and dons who wanted to think for themselves. Padua, in northern Italy, enjoying the protection of the 'Serene Republic of Venice' after 1222, became an international byword for student pandemonium, where, lacking a great Churchman to keep them in order, the students of the university virtually appointed and sacked their own professors and fought amongst themselves. Purgatory-induced educational endowments, a Europe-wide population increase, and new wealth generated from trade, Crusader lands, and improved agricultural efficiency, had suddenly unleashed a new economic and political energy which in turn provided the necessary resources from which the intellectual renaissance could take off.

In the Latin-speaking and Latin-reading international academic community, moreover, students and dons with radical interpretations of Aristotle, or of the nature of the Eucharist, travelled as they saw fit. Irishmen studied in France, Spaniards in Italy, Italians in Oxford, and so on. Medieval Europe's greatest experimental physicist, the Franciscan friar Roger Bacon (*c.*1214–1292), studied in Oxford under the great Robert Grosseteste, was arrested and imprisoned for his radical political beliefs in 1277, and over the course of his long life travelled between some of the great universities of the age, dying in Oxford on 11 June 1292.

But what exactly was the curriculum within which those radical new ideas could be shaped and expressed? It was based upon the Seven Liberal Arts, or Sciences (from the Latin *scientia* – 'organised knowledge'), that had first been developed by the Latin grammarians Donatus and Priscian and other late Roman teachers, though these sevenfold categories possessed sufficient flexibility to enable them to absorb and interpret much of the knowledge that came in via the new translations of books. These 'sciences' were split up into two groups of organised thinking: the *Trivium* ('three paths') and the *Quadrivium* ('four paths') – all of which were concerned with reason and with measured understanding. The *Trivium* consisted of grammar, rhetoric and logic – the three disciplines pertaining to spoken knowledge. For without grammar one would lack a common language, and without language one would not be able to use speech to effect, to produce rhetoric. And without logic, our language, speech and ideas would have no more intellectual direction than would the jumbled grunting of the beasts. The *Trivium*, therefore, taught clarity of thought and expression, and the ability to construct connected arguments.

The *Quadrivium* in turn taught *ratio*, method, order, or measurement. Geometry taught the eternal truths of proportion, as seen in the angles of divided circles and triangles. Arithmetic, likewise, aimed at opening up the mind to far more than simple counting, but rather to the ineffable wonder of things that never changed. For numbers never change, and have wonderful relationships with each other. Two and two always make four, and square roots never alter. Within the *Quadrivium*, music was not a performance art, but a study of eternal ratios. Pythagoras had first explored the mathematics of harmony, and Boethius in turn had written on Pythagorean theory and the proportions into which a vibrating string or blown pipe had to be divided to produce the notes of a scale. These harmonic ratios, moreover, not only produced elegant arithmetic sequences; they also produced chords and discords which were delightful or uncomfortable to the human ear. Surely these sequences must have been designed by God, in the respect that sweet chords, whole numbers and perfect geometrical shapes are naturally pleasing to us, as presumably they are to God Himself. Discernment, therefore, in speech, logic, music, or shape, was not a thing to be left to the individual. It was, rather, an absolute, divine standard, the general awareness of which lay implanted within the human psyche, yet which still needed education and intellectual discipline to raise to conscious understanding.

Astronomy, however, was the component of the *Quadrivium* which had practical applications not only in the devising of calendars, but also to the very understanding of the construction of the cosmos. A medieval undergraduate would learn that the Earth was a sphere, the surface of which, according to the 'Apocryphal' Old Testament book of *Esdras* (II:6:42), was believed to consist of six parts of land to one part of water. It consisted of the four elements of Aristotle, and around it rotated nine spheres of transparent crystal, nestling inside each other like the skins of an onion. The Moon was deemed to mark the divide between the 'sublunary' earthly and 'superlunary' celestial regions, for its monthly phases showed it to have mutable characteristics, whereas the perfect predictability of these phases nonetheless indicated its perfection. Beyond the Moon, the next six spheres carried the planets: Mercury, Venus, the Sun, Mars, Jupiter and Saturn. Surrounding these seven transparent planetary spheres was the black sphere of the night sky which, carrying the stars, spun around in 24 hours to provide a spangled background against which to plot the nightly course of the planets. And last of all came the 'ninth sphere' – the *primum mobile*, or Prime Mover – which was a sort of great celestial flywheel which somehow controlled the rest. It was in astronomy that the *Quadrivium* achieved its aesthetic completion, with its implicit mathematical elegance, and with its natural incorporation of the nine orders of angels, each one residing upon its appropriate sphere. The celestial, the angelic, the intellectual and the aesthetic all came together to lift men's minds from earthly sin and passion to God's eternal perfection, via the gift of reason. One might argue that no educational vision has ever been grander or more exhilarating than this fusion of the divine with the earthly as exemplified in the medieval universities. Muslims held a similar view.

While translated Greek and Arabic works lay at the heart of much medieval astronomy – for they, after all, were held as embodying the basic structural truths of the Universe – western European scholars developed and expanded their sources. Indeed, one of the most successful astronomical textbooks ever written was produced to help expound the *Quadrivium*, and did the rounds of classrooms for more than four hundred years, so it would even have been familiar to the young Sir Isaac Newton. This was *De Sphaera Mundi* (*On the Sphere of the World*), written around 1240 by one John of the Holy Wood, or in the more familiar Latinisation of his name, Johannes de Sacrobosco. Sacrobosco, who was variously said to have been an Irishman, and also a native of Halifax, Yorkshire, taught astronomy at the Sorbonne in Paris. His *De Sphaera Mundi* was a digest of the main sections of Ptolemy's *Almagest* that described the basic Earth-centred cosmology and crystal spheres mentioned above. It was saying nothing new scientifically, although the proliferation of copies of 'Sacrobosco' – especially after the invention of moveable-type printing in the mid-fifteenth century – firmly gives the lie to the popular myth that the medieval Church somehow tried to suppress astronomy.

And at the same time, the astrolabe – first popularised in the West by Gerbert of Aurillac (who became Pope Silvester II in 999) – became widespread across Europe. Indeed, astrolabes bearing Latin inscriptions were being manufactured in many European countries, especially in Italy and Germany, by 1400, and several hundred still survive in museum collections in Britain, Europe and America. Merton College, Oxford, alone possesses two, which have been dated, from internal astronomical and design features, to around 1350 and 1390 respectively.

The astrolabe was a form of personal computer which enabled its owner to solve all kinds of problems in planetary and stellar astronomy (as was mentioned in Chapter 6). But in addition to the instrument's often being used in university teaching and, in medieval paintings, as a symbol for wisdom and learning, it is clear from Geoffrey Chaucer's writings that by around 1381 the astrolabe was sufficiently common in England to warrant his writing a 'user's manual'. Chaucer's *Treatise on the Astrolabe* was the first book to be written in the English language on the use of a complex scientific instrument. The 'Preface' tells us that he wrote it for 'Lyte[l] Lowys [Lewis] my sone', who though only ten years old already displayed 'evydences' of 'abilite to lerne sciences touching nombres and proporciouns'. Sadly, history remains silent upon the subsequent life and career of the clearly very bright 'Lytel Lowys', and one is left to assume that he died young.

Chaucer also tells us, in *The Miller's Tale*, that an astrolabe, along with musical instruments, was among the possessions of his idle and lecherous Oxford undergraduate Master Nicholas, who, while preparing to seduce his elderly and credulous landlord's pretty young wife, adorned his chamber with 'His Almageste, and bookes grete and smale, His Astrelabie, longynge for his art...' If Master Nicholas did, indeed, possess an astrolabe and a copy of Ptolemy, he was clearly not a *poor* clerk, for such items were expensive. Even so, their inclusion as incidental student artefacts in a fourteenth-century work of comedy suggests that they were already firmly at home in that society.

Also becoming familiar across Europe by Chaucer's time was the mechanisation of the astrolabe, whereby weight-driven gear trains were made to rotate the brass *rete*, or star-map, of the instrument to simulate the rising and setting of the stars. Indeed, one of the remarkable features of medieval European society was a love affair with machinery and technology that went far beyond anything in antiquity, Islam, or even the Orient. Beginning, perhaps, with the ingenious metal joints and iron manufacture so vital in the production of knightly armour and weapons, there flowed a cascade of inventions after about AD 1000. These included improved agricultural equipment such as heavy ploughs, water clocks, mechanical clocks that rang bells and turned astrolabes, the ship's rudder, glass windows, windmills, vastly better watermills capable of doing heavy industrial work such as operating heavy forging hammers, crossbows, gunpowder, firearms, the widespread manufacture and use of the Chinese-invented magnetic compass, spectacles with lenses, complex locks and keys, and all manner of pulleys, gears and automata.

Of all of these devices, the most powerful in its analogical potential, and historically enduring, was the mechanical clock – that ingenious configuration of gears and weights whereby both Richard of Wallingford in Oxford and St Albans and Giovanni di Dondi in northern Italy first made the sky plates of an astrolabe rotate on their own and replicate in miniature here on Earth what God did with the spheres of heaven.

Attempts to measure the hours of the day and night by some kind of artificial means was not new in the 1310s and 1320s, when Richard of Wallingford was active. The Egyptians, Romans, and others, had used various types of water clock. In the tenth century AD, King Alfred of England had developed a burning-candle clock; while around 850, Pacificus, Archdeacon of Verona, supposedly made a water clock that could have been based on accounts of ancient Roman devices. Yet what happened around 1300 was radically different. For someone – probably either an Englishman, or an Italian, or else both simultaneously – found a way of using the meshing of gear ratios to mark out even units of time. A heavy weight was made to turn a 'race' of gears. But instead of the gears simply spinning out of control until the weight hit the floor, this new device contained an 'escapement', or a rocking mechanism, that released the energy of the falling weight in regular tiny beats or 'ticks'. And once the mechanical principle of this anonymously-invented escapement came to be understood, it spread like wildfire across Europe, so that by the time of Chaucer's death in 1400, *horologia* (Greek – 'to measure hours'), or clocks, had become commonplace.

Clocks were used to ring the bells that marked the 'canonical' hours that summoned the monks to their devotions – as was the still working clock in Salisbury Cathedral that dates from 1386. Soon afterwards, they were adapted to turn fingers around a dial, and then to replicate the phases of the lunar month, as in Exeter Cathedral; or to produce public spectacles in which wooden figures were made to move or tunes to play automatically on carillons of bells, as at Wells or Strassburg; or to turn the plates of an astrolabe, as in St Albans and Florence.

The mechanical clock represented something far more potent in mankind's developing awareness of both the heavens and the divine than a seemingly over-elaborate way of finding the time on a cloudy day. It reflected that mind which had created the very firmament itself, indicating that it could not only be glimpsed by the mortal creation, but was sufficiently present in that mortal creation to enable human beings to replicate the very motions of that firmament here on Earth. The clock, in short, was a spiritually potent invention, and it is not for nothing that in fourteenth- and fifteenth-century illuminated manuscripts clocks were often shown as being attended to or adjusted by angels and the Virgin Mary, as a way of symbolising the divine governance of the cosmos.

Just as the mechanical clock formed one analogical parallel between mankind, the heavens and God, so the logically systematic and physically ingenious scholars of medieval Europe discovered others that were to be equally far-reaching. One of these arose from their further development of the earlier researches described in the newly translated optical writings of Aristotle, Ptolemy and Alhazen.

Light fascinated medieval scholars – Jewish, Christian, and Muslim alike. For in their monotheistic Universe, where human beings enjoyed a unique, albeit sin-tainted, relationship with God, light was seen as that wonderful natural agent which originated in God's heaven, passed through the sphere of the Moon, and filled with radiance the world in which we live. For at the creation as recounted in *Genesis*, God had separated the light from the *tangible* darkness, or chaos; St John's Gospel spoke of Christ as the 'Light of men' and of the world, while Alhazen in Cairo had spoken of light's divine majesty. And just as religious awe played a significant part in motivating Alhazen's optical researches, so it did for medieval Europeans.

One of the first and most historically significant of these was the Englishman Robert Grosseteste, though it was his pupil Roger Bacon who, during his long and sometimes troubled itinerant career, developed his mentor's ideas, to make some of the first great advances beyond the discoveries described in Alhazen's *Optical Thesaurus*. Learning from Alhazen about reflection, refraction, physiological optics and the nature of meteorological distortion, Bacon went on to perform a brilliant series of observations and experiments on light. An optical phenomenon which had puzzled ancient Greeks and north Europeans, along with Arabs such as Alhazen, was the nature of the rainbow. However, rainbows are much more common in the damp climates of Europe than in the generally dry air of Cairo, which may go towards explaining the growing body of work on the optics of the rainbow in medieval Europe.

While Aristotle had produced an early physical model for the rainbow in his *Meteorologica* around 350 BC, and his work in turn had been amplified by that of Alhazen, these explanations, based as they were on the reflection and refraction of light within a cloud, left many things unanswered. As an instinctive research physicist, however, Bacon began to experiment in his 'laboratory' by studying the colours produced when light was refracted by simple lenses or water droplets, in much the same way that Alhazen had when he observed the primary and secondary spectra produced by a suspended transparent sphere.

Then, building upon Aristotle's, Alhazen's and Grosseteste's foundation, Bacon realised that a rainbow was, in fact, but an arc, or part of what was theoretically a full ring of light, most of which, of course, was below the horizon. That ring, moreover, formed the base of an invisible cone, the point of which was occupied by the Sun, and the central axis – from the Sun-point to the centre of the base ring of light – passed through the line of sight of the person looking at the rainbow. Whether a particular person sees a rainbow when the Sun comes out after rain, therefore, depends on where he or she is standing with relation to the Sun (at their back), the cloud (to their front), and the height of the Sun and cloud in the sky.

One can explain, therefore, why we never see a rainbow at noon in June, because then the Sun is too high in the sky for a bow to form, for its theoretical cone would scarcely be noticeable above the northern horizon. And likewise, one can explain why the most spectacular bows are at morning and evening, for then the Sun is on the horizon, and the cone is so high in the sky as to produce what can be seen as a semicircular rainbow.

But Bacon did not stop here. In his attempt to understand the optical geometry of the bows, he then went on to study those artificial or 'laboratory' rainbows which result from the spray thrown into the air by water wheels when the Sun is in the right position. (Such artificial bows are most familiar to us today when produced by garden lawn sprinklers which generate small clouds of moisture.) From all of his observations and experiments, and after making careful measurements with an astrolabe, Bacon came to the conclusion that when a rainbow occurs, the light rays entering our eyes emerge from the cloud of moisture at a fixed angle to that at which the direct sunlight entered the cloud. This angle he cited as being 42 degrees of arc – a value which (although slightly more refined) is still accepted today.

In the wake of Bacon's researches, Theodoric (sometimes called Dietrich) of Freiburg further pioneered the experimental study of the rainbow. Theodoric realised that central to the problem was the behaviour of light rays in the *individual* droplets of water which composed the coloured bow. Like Bacon, he studied the behaviour of light under controlled laboratory conditions, and used a hollow glass sphere (probably a physician's flask used to examine a patient's urine) filled with clear water, as an experimental raindrop. Theodoric then admitted controlled rays of bright sunlight into the flask within a shaded room, and found the familiar pattern of colours coming out, which was attributed to the 'corruption' of the white light. But very significantly, Theodoric noticed that inside the sphere, the light was both *refracted* by the water and *reflected* from the concave inner surface of the sphere, or droplet. Only after these compound refractions and reflections did the light leave the water. And what was more, the paths which they followed were in accordance with exact geometrical ratios.

Experimental physics, one might therefore argue, was born in the laboratories, or work rooms, of a succession of predominantly English and German clerical scientists in the thirteenth and early fourteenth centuries. Beginning with the observations of Aristotle and Alhazen, Albertus Magnus, Robert Grosseteste, Witelo of Silesia, Roger Bacon, John Pecham, and Theodoric of Freiburg studied the anatomical structure of the perceiving eye, measured the angles existing between the Sun, themselves, and the rainbow, and then went on to abstract specific pieces of optical phenomena, such as water droplets, for special and intensive study. Using flasks of water to investigate internal reflection and refraction under controlled conditions, they came to develop a physical 'model' for the rainbow that could be further refined and tested against improving experimental results that extended down to the prism experiments of Isaac Newton in the 1660s. For here, one is a million miles away from the middle ages of popular legend, and from those hoary tales which encourage us to believe that medieval people thought rainbows sprang up from pots of gold.

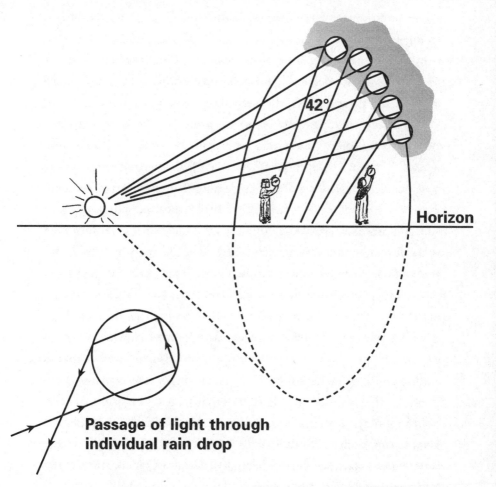

**Passage of light through
individual rain drop**

The rainbow as it was understood by 1400. The
bow was part of an optical cone, with the Sun
at its apex. Millions of individual droplets of
water in a cloud both refracted and internally
reflected the sunlight that passed into them
[inset]. The light forming the bow came out
of the cloud at an angle of 42 degrees to the
sunlight going in, as could be measured with
an astrolabe. Two observers standing on the
same axial line of the optical cone, moreover,
would see two slightly different rainbows.

In addition to its fascination with explaining the Creation and tracing God's hand in cosmology, optics and mathematics, and even with attempting to replicate that creation in the astrolabe and the mechanical clock, medieval Europe's most enduring analogy of the divine on Earth was the Gothic cathedral.

While big buildings were by no means new in the twelfth century, and north European architects were just becoming familiar with Marcus Vitruvius' first century AD *Ten Books of Architecture* – the most detailed treatise on architecture and civil engineering to come out of the Roman world – something very radical now took place.

Undoubtedly, crusading contacts with the Muslim world played a part, for the pointed arch, which had been invented by Arab architects and had been hitherto unknown in the West, became an integral feature of the emerging Gothic style. Yet what caused this new style to suddenly blossom forth, first at St Denis and Chartres in northern France in the early twelfth century, and not only to reach technical maturity within a very brief span of years, but also to be envisaged as a 'microcosm' or image of the Universe, is difficult to explain. Quite simply, Gothic architecture was part of that same cultural and spiritual explosion which brought into being the universities, new orders of monasticism such as St Bernard's Cistercian Order, and a whole movement of spiritual and intellectual renewal which constituted the renaissance of the twelfth century.

Europe's ancestral architecture had evolved within cultures which enjoyed the high Sun and the bright skies of the Mediterranean: from Greece and Rome, where the strong light and summer heat favoured small windows and thick walls. After its famous pillars and cross-lintels, classical architecture's fundamental load-bearing structure had been the semicircular arch, locked in place with a keystone at the top. Yet the semicircular arch needed massively thick walls of brick or stone if the arches were to support a great vaulted roof, such as in a pagan basilica or great Roman church. And such load-bearing walls would permit only small windows. In a Mediterranean climate, this was not a particular disadvantage; but at the latitudes of Paris, Cologne, Salisbury or York it was a distinct disadvantage, because the resulting interiors rarely rose above sepulchral gloom.

The Gothic arch, however, has very different load-bearing properties. When great weight is placed upon it, the it tends to burst outwards at the point where the upright supporting columns and the curved planes of the arch meet. Yet if – as an anonymous early twelfth-century French master mason discovered – a second semi-arch is made to lean against this potential bursting point and apply external pressure at the critical spot, then the main arch will remain standing in spite of having to bear an enormous weight above it. These semi-arches came to be called 'flying buttresses' – quite literally, because they seemed to hang in the air – and acted as finely balanced stone girders that channeled the weight of the superstructure into the Earth.

As a result, the arrangement of flying buttresses around a great building, such as a cathedral, allowed the introduction of two daring structural innovations. Firstly, because the carefully positioned flying buttress took the massive weight of the roof and tower structures, and filtered them into the Earth through a mathematically arranged filigree of stone girders, the walls no longer had the primary job of supporting the superstructure. Instead, they carried a much smaller part of the building's weight, in consequence of which the walls could now be penetrated with enormous glass windows. Secondly, because the arches were pointed they were able to impart an interior geometry to the building, which seemed to rise up to a great height and create a breathtaking sense of soaring space. The genius of Gothic, therefore, lay in its ability to create vast interior spaces that were flooded with light.

Indeed, one has only to look at a great Gothic church such as Chartres, Salisbury, St Chapelle, Windsor, and many others, to see this effect. Their walls seem to contain almost as much glass as they

Section of a Gothic cathedral.

A. The flying buttress, and heavy masonry buttresses, at right-angles to the building, allowed the weight of the superstructure to be channelled away from the walls.

B. Plan of the building from above. Note the vertical load-bearing structures, shaded, allowing vast windows. The dotted lines show the arrangement of pointed arches that held up the roof and tower.

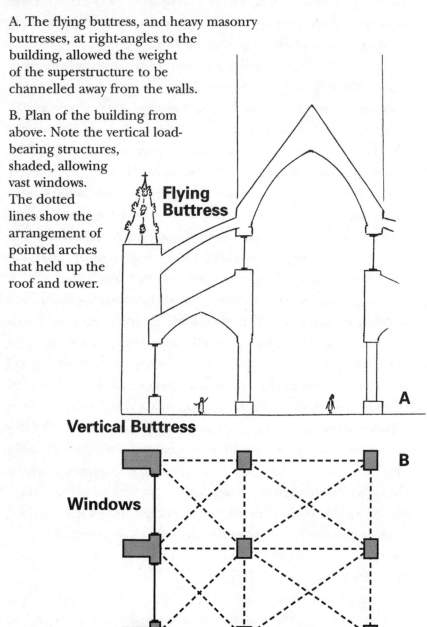

Flying Buttress

Vertical Buttress

A

Windows

B

do stone, and when you take in the proportions with your eye, and ask 'How is all that weight of roof and tower supported by such apparently flimsy glass-filled walls?', the answer lies in the flying buttresses and arches, which result in a building rather resembling a house of cards: it stays up because every downward thrust of weight is matched with an exact support, so that the whole vast edifice is in a permanent state of equipoise. Not only were the Gothic cathedrals of northern Europe, with their spires and towers, the world's first skyscrapers; they also resembled modern skyscrapers insofar as they stayed up by balancing forces through a filament support structure that virtually dispensed with the need for walls. Quite simply, precision engineering replaced brute mass, and these great religious edifices can be thought of as static machines that are perfectly designed to handle colossal vertical forces which stabilise them into the Earth. Indeed, it was highly appropriate that the Gothic cathedrals and abbeys should also have housed the first mechanical clocks.

Few architectural styles, moreover, have been driven so close to the limit as has Gothic, for once the basic load-bearing geometry of the pointed arch had entered into the minds of medieval architects, then they pushed and pushed. How high could you make the tower while still seeming to create a vast open space beneath it? How many square yards of glass window could you safely incorporate? And how daringly could you shape your arches and still not have your building fall down?

These technical innovations were reflected in the changing styles of the language of Gothic architecture, thereby producing the Early English, 'Decorated', and the French 'Flamboyant' styles. But sometimes they went just a little too far: stone support columns would start to bend or develop disturbing cracks as tens of thousands of tons of stone, timber and lead seemed to hover above the glazed walls, and at Salisbury and Wells in particular, later masons had to build ingeniously concealed additional supports to prevent the cathedrals from collapsing.

Yet why, one might ask, did the medieval Church spend such vast sums of money in building these cathedrals? The real answer lies in their symbolic power, for they were microcosms, or little models, of the whole cosmos. For one thing, they were profoundly geometrical buildings, in which the classical proportions of the cube, double cube, circle and conic section were all brought together to form a spectacular aesthetic whole. Indeed, one reason why these buildings still make modern visitors gasp is because they embody classical proportions and have an inner symmetry which delights the eye and humbles the mind, in very much the same way as does a brilliant starry sky. This geometry of awe, moreover, was often supplemented by a geometry of detail and love of puzzles, such as the great circular maze of black and white stones that is laid into the floor of Chartres Cathedral, and which compels the observer to rack his or her brains trying to fathom a way through it. For just as God clearly enjoys setting intellectual puzzles in nature – such as the movements of the heavens, or the causes of the tides – so one complements this ingenuity in architectural terms by filling God's house equally full of mathematical delights.

When one remembers, furthermore, that those vast buildings also housed mechanical clocks, that their clergy needed sufficient astronomical knowledge to calculate the date of Easter, and that their affiliated schools taught the use of the astrolabe and Ptolemy's epicycles, one develops an appreciation of how the cosmos was seen to be replicated, in its imperfect human way, here on Earth. The widespread belief that the medieval Church attempted to suppress astronomical knowledge is proved to be ridiculous by the sheer weight of evidence to the contrary.

At this point it would also be appropriate to correct another widespread myth surrounding the Gothic cathedrals: namely, that they were almost spontaneously constructed by ecstatic monks and peasants in the grip of religious zeal. While religious zeal and heavenly vision were undoubtedly powerful stimuli, Europe's cathedrals were – as one might expect from such daring and complex structures – the products of a highly-trained elite corps of professional architects, master masons and skilled workers. The surviving records of the English cathedrals alone, with their often named builders and contractors, and even sometimes wage accounts, provide ample substantiation of the fact.

Yet if one really wants to enter the world of an early thirteenth-century master mason, one has no finer source than the drawing-book fragments of the Frenchman Villard de Honnecourt from the 1230s, which are now preserved in the Bibliothèque Nationale in Paris. Like most men in his profession, Villard was employed internationally, and the surviving thirty-three vellum sheets of his notebook, written on both sides (there had clearly been more, but sadly they have been lost) abound with drawings of new buildings, buildings under construction, cranes, hoists, water-power saws for cutting up tree trunks, hydraulic devices, perpetual motion machines, dangerously flimsy-looking scaffolding, and 'pre-clockwork' devices that attempted to tell the time by means of falling weights. Indeed, what Villard's notebook tells us is how highly organised the building trade was by his day, and what an abundance of actual and suggested technology existed. The notebook also tells us that even 800 years ago, at least some people were willing to let their imaginations run to the odd touches of 'fantasy technology', as was Villard when making a drawing of a wheel with seven hammers carefully pivoted on its circumference. Once the wheel had been set in motion, the changing centres of gravity caused by the rising and falling hammers meant that the wheel should have kept running for ever!

Late thirteenth-century and fourteenth-century European astronomy was in many ways indebted to Re-Conquista Spain, for, as we have seen above, the capture of Islamic Toledo in 1085 released a whole wave of translation into Latin of classical and original Arabic scientific books. For example, the eleventh-century *Toledan Tables* of the Spanish Muslim astronomer Al-Zargalli (*c.*1070), with their Ptolemaic and Arabic astronomical data, became the prototype for the Latin *Alphonsine Tables*, variously compiled, it has been suggested, by Spanish Christian astronomers and by the Frenchman Jean de Linières in honour of King Alphonse of Castile around 1272, and which had a major effect on practical and computational astronomy across Europe. Curiously enough, though, there was a major difference between the development of astronomy in medieval Europe and in the Arabic world, and that concerned original observation of the heavens. As we have seen (in Chapter 6), the Arabic astronomers came to build observatories across the Muslim world, from Samarkand to Spain, and they used these observatories to remake Ptolemy's observations using large and accurate instruments, and to conduct elaborate analyses of the movements of the planets.

In Christian Europe, by contrast, no *research* observations of the heavens seem to have been made, for while the astrolabe was in widespread use for calendrical, time-finding, and similar practical activities, no European astronomer built a research observatory from which to perform original investigations until that of Bernhard Walther, which was active in Nuremburg between 1475 and 1504. Exactly why this was the case is not clear, but it might have had something to do with the fact that from the eighth century onwards, Arabic science contained a large dimension of hands-on practice, whereas that of Europe, in spite of experimentalists such as Roger Bacon and Witelo, science tended to be dominated by philosophical issues.

These philosophical issues, however, imparted an energy to medieval cosmology which, far from being backward-looking, often makes it seem startlingly modern. The main contributory factors were those two elements which we have seen influencing culture so many times already: Aristotle and Christianity. Aristotle was cosmologically influential through his argument that a First Cause was logically necessary to set all of the other causes of the Universe in motion, although his concern with observed cycles in nature made it impossible to know when or why that First Cause had acted. Christianity, on the other hand, was about a personified First Cause – God – who worked directionally and not cyclically; and because He had existed before the world was made, and would, according to Scripture, continue to exist after it had passed away, He was by definition infinite.

Twelfth- and thirteenth-century scholars were, of course, familiar with St Augustine's arguments about God pre-existing time and nature (for time was marked by the motions of created heavenly bodies); yet they asked, could there be any sense of time in heaven? Presumably not, for the saints and the saved and blessed dead are in a state of joy in the immediate company of God with which they can never tire or grow bored. For boredom hinges on a sense of time, as one remembers past joys and looks forward to future joys. But can one have a more exquisite state of existence than being in God's presence? And can one look forward to any greater joys? Of course not. Ergo, time cannot exist in heaven – only an eternally wonderful present.

Conversely, hell must also exist outside time, and possess a set of physical rules all of its own. How can one suffer eternally without in some way growing accustomed to the pain, and in consequence, not suffering quite so much? And why, if fiends and furnaces are perpetually sawing, searing and roasting the flesh of sinners, is that flesh never destroyed, but always ready to suffer anew? In fact, what we are encountering in medieval notions about creation, heaven and hell are the first discussions of what might be called relativistic ideas, in which the normal laws of time, space and matter do strange things! Classical pagan thought had nothing quite like it, and while Islamic philosophers had a strong sense of heaven as Allah's eternal and wonderful space which knew only joy, it was in Christian Europe that these relativistic ideas came especially to be explored.

Several medieval scholars discussed the possible implications of infinite space and relative time, and one could perhaps argue that modern theoretical physics can trace its genesis to fourteenth-century Oxford and Paris. Lying at the heart of this intellectual quest were growing doubts regarding the correctness of Aristotle's laws of motion, which then formed the bedrock upon which medieval physics was built. At the bottom of Aristotle's physical science was the concept of absolute space, time and motion, along with the central idea that motion was an innate condition of all things. Yet what if rest were a natural condition of bodies, and they moved only when some force or *impetus* acted upon them, and the resulting motion was directly relative to the impetus? One encounters here the seeds of the subsequent physics of inertia – a concept that was really contrary to the ideas of Aristotle, and yet three centuries later would become fundamental to the gravitational physics of Newton. And without Newton, modern physics – including that of Einsteinian relativity and quantum theory – could not exist.

The early fourteenth-century Oxford scholar William of Ockham was one of the first to attack Aristotle's ideas, arguing that when a body 'moved', this motion was not necessarily an absolute in itself, but was merely *relative* to the other objects in the space through which it was passing. After 1340, the idea was taken up, explored and developed by Thomas Bradwardine (who became Archbishop of Canterbury), by Jean Buridan, Rector of the University of Paris, by Buridan's pupil Nicole de Oresme, the future bishop of Lisieux, and many others. From their studies of motion, these men went on to develop revolutionary ideas about the coordinate geometry of space, and how that space was defined not by any absolutes, but from the relative positions and motions of the objects that occupied it. In many ways their researches were what might be regarded as logically controlled 'thought' experiments, but they did indeed consider the effects of observable physical phenomena, such as the motion of arrows flying relative to the wind, and the relationship between objects free-falling from the masts of moving ships and the surrounding ocean and land.

All manner of cosmological possibilities came to be explored by these fourteenth-century clerical scientists. For example, William of Ockham (who probably died of the Black Death of 1349) defined that famous logical rule still called 'Ockham's Razor': namely, that the simplest logically correct explanation for a given problem is more likely to be true than is a very complex explanation. And Bradwardine argued that there was no logical reason why an infinite Creator God could not have made an infinite Universe rather than a Universe bounded by the classical spheres.

But it was in the writings of the fifteenth-century German cardinal, Nicholas of Cusa (Kues), that these ideas reached their fullest development. Cardinal Nicholas came to argue against the Aristotelian distinction between the physics of the terrestrial and celestial realms: if all motion was relative, then the same forces that acted upon a flying arrow must also act upon the planets. He also argued that astronomical bodies were not made of a mysterious 'fifth element' but that they must, like the Earth, be made of earth, water, air and fire, to provide the cosmos with a uniform physical composition. And he realised that as the cosmos was the work of an infinitely wise and all-powerful Creator, it was at least possible to consider that cosmos as being infinite, so that the Earth itself might be no more than one of many bodies moving relative to each other in a Universe that had no natural centre, but only extension in all directions.

It is impossible to overestimate the impact of these ideas upon the subsequent history of astronomy and physics. Their criticisms of Aristotelian doctrines of motion were to act as starting points for seventeenth-century figures such as Galileo Galilei, René Descartes, Christiaan Huygens and Isaac Newton, as were, to a lesser extent, their speculations upon the possibility of an infinite as opposed to a bounded Universe. Similarly, their ideas on the geometry and logic of space provided a key concept that would reach its full experimental and mathematical realisation after 1600: namely, that one set of intellectual principles runs through the whole creation – terrestrial and astronomical. In turn, Nicholas Copernicus was familiar with Nicole de Oresme's ingenious suggestion – which in itself was an application of Ockham's Razor – that it was more economical for a Creator God to make the Earth rotate on its axis every 24 hours than to make the whole vast cosmos spin around a stationary Earth.

This achievement, moreover, did not come, as is popularly supposed, from daring, quasi-secular free spirits who lived in fear of the Church. For, quite to the contrary, not only were all of these men priests, but priests of the highest seniority: heads of universities, bishops, archbishops and cardinals. None of them, furthermore, were punished for their thoughts, and all died in their beds. Their criticisms were levelled at the dogmatic acceptance by their academic contemporaries of the ideas of a classical Greek philosopher, and not at religious belief, as the late medieval Church possessed a remarkable intellectual expansiveness on matters of science.

To see medieval Europe, therefore, as dark, intellectually backward, and hostile to scientific ways of thinking, is false. Although the intellectual life of the time was undoubtedly dominated by the Church and its clergy, it was by no means true that these men, and a tiny handful of women such as the twelfth-century Reverend Mother Hildegard of Bingen, who were interested in the natural world, were closed-minded, bigoted or persecuted. As a spiritual and learned corporate body, the medieval Church spearheaded education, translation, classical studies, constitutional politics, and a sustained insistence that, as human beings were rational creatures made in the image of God, reason provided the razor that could cut through the knots of pagan superstition, and lift us up to the divine. While intellect was not necessary for individual salvation, God nonetheless had given rational minds to humanity as a divine species, so that the brightest and the cleverest could help clear the path that their simpler brothers and sisters could follow to Christ. At the very end of the intellectual quest, our reason was sufficiently humbling in itself to remind even the cleverest bishop or scholar of his need to make a final leap of *faith* as the last act of the human intellect, as he transcended reason and humbled himself at the feet of God. Yet this faith was in no way seen as blind or superstitious, but rather as the fulfilment of a promise made by a Creator whose gift of reason told us that he could be trusted.

The medieval cosmos, moreover, was the very antithesis of the minimalist. In its exotic intermixing of the spiritual and the physical, the rational and the transcendent, the finite and the infinite, and its successful incorporation of the best of pagan classicism together with the incarnation on Earth of the Creator God of *Genesis*, it constituted one of the most dynamic and far-reaching developments in the history of human thought. Quite simply, without those Angels Ascending the Spheres, so much of the content and metaphor of modern science and civilisation would not have existed.

N I N E

From the ashes of the Apocalypse

One of the ironies of the medieval cultural achievement, in science as in everything else, was its assumed provisional or temporary character. For while there was a strong sense of historical direction, one key component of modern culture was missing from the medieval world view: the concept of progress and improvement. Instead, some late medieval scholars and prophetic visionaries, just like their early Christian forebears of a thousand years earlier, tended to believe that the Apocalypse was imminent and that the cosmos as they knew it would soon be 'rolled up like a curtain', and conventional time and space would cease to exist. And while, one might argue, their 1,400-year or so wait for Armageddon should have made them somewhat blasé, there were plenty of things going wrong within their world to convince Europeans that the Last Days were upon them at last.

For one thing, the Crusaders had already been driven out of the Holy Land by new and dynamic Muslim leaders. The fearful Black Death, or bubonic plague, which had broken out for the first time in Europe in 1347, had rapidly become endemic, and within a century had considerably reduced the population and economic resources of

Christendom. Further new and terrifying diseases such as the Sweating Sickness and – most frightful of all – syphilis, had suddenly appeared to scourge Europe just before 1500. All of these added to the sense of divine wrath and impending doom. Surviving records, moreover, mention the appearance of more comets, blood-red aurorae borealis, mock suns and other doom warnings than before – although these could well have been the results of increasingly precise scientific observation.

But most frightening of all was the lightning ascendancy of that Near-Eastern power which Europeans often referred to under the generic title of 'The Turk'. As the Christian Greek Orthodox Byzantine Empire gradually waned in power after their fellow-Christian Venetians sacked Constantinople in 1204, so the Ottoman Turks grew from strength to strength as Islam regained the crusading initiative. Little by little, this radical Islamic power gobbled up the Byzantine Empire during the course of a century, until the dynamic young Sultan Mohammed II struck the *coup de grace* in 1453. In the spring of that year, he laid siege to Christian Constantinople, and after an heroic defence by the Greeks, in which the last Byzantine Emperor, Constantine XI, was killed defending his capital city, Mohammed II entered Constantinople in triumph on the afternoon of 29 May. With the fall of Byzantium, that Christian state which western European Roman Catholics saw as a buffer between themselves and the increasingly militant Islamic Near East, shock waves of fear resounded through Vienna, Rome, Paris and London.

Now that Mohammed had conquered Constantinople and was occupying the European Balkan states, what could keep him out of Austria, Italy and France? However, because of internal political problems within the Ottoman Empire (which were not understood in Europe in 1453), and also the death of Mohammed II in 1481, the dreaded invasion of western Europe never came. Even so, during the next century or two, several of Mohammed's successors would take predatory bites at the European cherry, including the seizure of Belgrade in 1521, the annexation of Hungary five years later, and a full-scale but fortunately failed siege of Vienna in 1529, along with the Turkish navy's constant harassment of south European ports and shipping. While the growth of European military, technological and economic strength had turned the tables against any serious attempt at a European invasion by 1650, in 1460 no-one in Paris or Florence could read the future. Sadly, the Turkish occupation of the Byzantine world left a legacy which is still very much with us today, for the ancient simmering hostilities between Christians and Muslims and their cultural traditions lie at the heart of the Bosnian, Kosovan and other Balkan atrocities.

With all this going on, therefore, it is hardly surprising that Catholic western Europe believed that Armageddon was almost upon it, while the overrun Byzantine Orthodox Greeks thought that it had already arrived. Medieval European civilisation was seen in some quarters as drawing to a close, as all sorts of people – from wandering soothsayers to men of learning – became acutely aware of old prophecies that foretold the end of time and the final act of chastisement for a sinful world.

For one thing, an old prophecy said that the Byzantine Empire would end as it had begun – with a reigning Emperor named Constantine whose mother's name was Helena. And while the royal house of Palaiologos had, over its last few decades, been careful to Christen its heirs with other names, it seemed ironic that Prince Constantine found himself Emperor in 1448, following the prior and unexpected deaths of his elder brothers John VII, Theodore II and Andronicus. And the mother of the imperial princes was named Helena!

Two of the not entirely orthodox passions of this age were visionary prophecies and numerology, whereby the end of time could be reckoned by interpreting the mystical significance of climactic events and of numbers (as was mentioned at the end of Chapter 2).

Although Jesus Christ himself had quite unequivocally told his disciples in the Gospels that no human being could ever discover when He would return to bring about the Last Judgment, this did not stop the growth of an active heretical tradition within Christianity which attempted to determine exactly when the end of time would be.

There were several Biblical texts which, it was believed, contained the key. Had not God taken *six* days, as recorded in *Genesis*, to create the Universe? And did not *Psalm* 90 state 'One Thousand Ages in thy sight are but as Yesterday'? If, therefore, the numbers 6 and 1,000 were to be multiplied together, it could be argued that God intended the Creation to last for 6,000 years, so that if its age upon a particular date was known, then the time remaining could be calculated. The answer to this question was also believed to lie in the Bible, for if the successive generations of the Children of Israel, recorded in the Old Testament, were to be added up, then when these were integrated with the St Matthew's Gospel genealogy of Joseph – who had married Mary, and therefore become the worldly stepfather of Christ – it would be possible to calculate the age of the world when Christ was born. And if this figure could be determined, then simple arithmetic was all that was required to establish how many of the 6,000 years remained.

The Christian scholar Julius Africanus had pointed the way around AD 217, when in his *History of the World* he calculated that 5,500 years had already elapsed when Christ was born. During the late medieval centuries, several other dates were produced, such as those extracted from the canonical Old Testament book of *Daniel*, from its apocryphal Biblical books of *Esdras* and *Enoch*, and from elsewhere, not to mention such significators as were believed to lie embedded in the *Tiburtina* and the Roman Sibylline prophecies. And so did not the contemporary disasters of plague and famine, the fall of Byzantium, the imagined imminent collapse of Christian Europe, the horrors of war now being fought with gunpowder, and the fleeing of homeless refugees out of the east, all mark the arrival of those terrors which Christ himself said would precede the end of time?

The year 1500 seemed pregnant with doom. The number 5 had all sorts of significances to numerologists; and when it was multiplied by three – the number of the Holy Trinity – the result was 1500. Some people said that there was another good reason to expect the Second Coming of Christ and Armageddon in 1500: the conduct of the Pope himself. Roderigo Borgia – an aristocratic Spanish cardinal, who became Pope Alexander VI in 1492 – was a moral disgrace who had several illegitimate children (officially nieces and nephews), the best known of whom were his psychopathic son Cesaré (whom Alexander made a Cardinal while Cesaré was still a teenager) and daughter Lucrezia, who came down through history (quite unjustifiably, in fact) as a notorious poisoner. Surely the end of time must be in the offing when such a man had the historic title of Christ's Vicar on Earth and successor to St Peter. Alexander proclaimed the year 1500 to be a 'Jubilee Year' – partly as a way of diverting popular fears about impending Armageddon, but also as a way of raising additional revenues from the sale of Indulgences to worried sinners visiting Rome. But 1500 passed without serious incident, and the world continued to exist.

It is not without irony, therefore, that in the midst of what contemporaries saw as God's imminent destruction of the world, that movement which we now call the Renaissance flowered. It flowered firstly in Italy, and did so for a variety of factors. For one thing, the Italian States – especially the central and northern States of the Papal dominions, Tuscany, Venice, Genoa and Milan – contained the richest and most populous cities in Europe. And as the historic heartland of the ancient Roman Empire, Italy possessed a culture and sophistication which had not only survived the onslaught of a succession of barbarian invasions, but which continued to thrive in spite of its contemporary political instability. (It was not for nothing, indeed, that Niccolo Machiavelli wrote what scholars consider to be the first study of scientific realpolitik, *The Prince* (1513), at this time, in the wake of the French invasions of Italy in 1494 and the subsequent (1527) sacking of Rome by the Germans.)

However, another major force behind the Italian Renaissance was the influx into the West of Greek scholars, books, and other intellectual treasures. Yet while that influx greatly swelled after Christian scholars fled to the West after 1453, it should not be forgotten that for the sixty years prior to the fall of Constantinople, Greek scholars had been coming to Italy and beseeching a succession of Popes for military assistance against the growing power of the Turks. In 1393, for instance, Manuel Chrysoloras had visited Italy on one such mission, and remained in western Europe as a representative of the Greek Orthodox Church. During his decades of residence in the West, Chrysoloras aroused interest in the classical Greek language amongst Europe's scholars, and while he was not concerned with the sciences, he is still considered the founding father of classical Greek learning in Renaissance Europe. For almost 1,000 years beforehand, however, Europe's Latinate learned culture had generally known the Greek authors only in translation – often via Arabic intermediaries, or in some cases through thirteenth-century translations directly from the Greek, such as those of William of Moerbeke. But now, Greek originals of Plato, Aristotle, Euclid, Ptolemy, and hundreds of others, entered Italy, and scholars were agog with excitement. Here were the foundation texts of classical European civilisation, not filtered through the minds and the pens of translators, but *a fontibus*, from the fountainhead, or their pure source. The classical Greek language began to be studied first in Italy, and then spread outwards into the rest of Europe's universities, to facilitate the reading of these texts. It was also fortunate, moreover, that this influx of original Greek material coincided with the invention of printing, which also took place in the 1450s, for printing presses commissioned Greek type fonts to facilitate the printing and wholesale distribution of these texts. After 1490, Aldus Manutius of Venice established what was to become a world-famous press – the Aldine Press – which specialised in printing accurate classical Greek and Latin texts, including editions aimed at the international student market.

When, therefore, one combines the invention of printing with the post-1453 influx of 'pure' Greek texts, one sees one of the most powerful mainsprings of the Renaissance come into action. Printed books were not only much cheaper to produce than were hand-written manuscript copies; they also possessed what modern scholars call 'typographical fixity', in that every student who possessed a copy of, let us say, the Aldine edition of Plato's *Republic* – be he in Bologna, St Andrews, or Cracow – knew that he was working with a proof-corrected and hopefully error-free standard source. Without this possibility, modern critical scholarship could never have got off the ground, for otherwise no-one could have been certain that his unique manuscript copy was not corrupted with copyists' errors.

Printing was also to be of especial benefit to the study of astronomy, science and medicine, insofar as it now became feasible to illustrate texts with correct and detailed pictures and diagrams. One of the problems with medieval manuscript books on scientific subjects was the uncertainty of knowing whether the copyist's drawings were correct, and manuscript editions survive in which serious diagrammatic errors have been made. But with a printed picture made either from a wood-block or copperplate engraving that had been proof-checked by the author before it had gone to press, one could be absolutely certain that the pictures made sense and keyed in with the text. Illustrated books containing astronomical drawings, pictures of instruments, botanical depictions, and architectural, engineering and technical engravings, came to abound after 1500, while Andreas Vesalius's *De Fabrica Humani Corporis* (*On the Construction of the Human Body*) of 1543 created the genre of the anatomical atlas, and his dissection plates still amaze one with their beauty and accuracy.

Indeed, it is impossible to overestimate the significance of the new Greek sources, conjoined with printing, in determining the direction of European civilisation, for like all great turning points, no-one at the time knew where they would lead. The fragmentation of a once-unified Roman Catholic Europe, which came in the wake of Martin Luther's Reformation, depended on printing for its rapid spread. For while it is true that Alexander VI, his warmongering successor Julius II (who was also Michelangelo's patron), and other aggrandising Popes were causing rumbles of serious concern across Europe, and most especially in Germany, Luther's radical theological stance depended in many ways on a free and easy access to Bibles not in Latin, but in the vernacular languages, which would have been physically impossible before printing.

At the heart of Luther's Protestant Reformation, especially as it developed in the 1520s, was a shift in the emphasis of Church authority. Instead of assenting to the Papal Succession, to the mediation of the sacraments from Christ to mankind via the Vatican establishment, and to the priest's ability to absolve men from their sins on the strength of the authority conferred on him by the Pope, Luther went back *ad fontes*, to the Bible, for spiritual truth. For, he argued, men and women need no complex Roman hierarchy through which to relate to God when they have the words of Christ himself in the New Testament. All that a person need do is study Christ's teachings on the printed page, reflect upon them, pray, repent, and have faith. No priests were needed, and the papacy came to be seen by the rapidly growing legions of predominantly north European Protestants as a perversion of Christ's teachings, and not a guide through them.

The only canonical or acceptable Bible in use at this time was the Latin Bible translated by St Jerome in the 380s AD. Yet while

Jerome's Vulgate Bible had been so named because it was in the vulgar, or popular, tongue of the Roman world of its day, the Latin language had, in the meantime, long since ceased to be the speech of ordinary folk, having become rather that of scholarship and of the learned professions. Luther, therefore, saw a vernacular Bible as imperative, and his monumental German Bible of 1522 thoroughly alarmed the Roman establishment. And as the ideas of Luther and other Protestant theologians such as Ulrich Zwingli, Jean Calvin and Thomas Cranmer spread from Germany into other countries, the Bible was soon being translated into French, English, Swedish, and other languages, and at the same time splitting Europe asunder as Catholics and Protestants began to wage war for territory and their respective perceived routes to heaven.

Quite apart from its theological and political consequences, the Reformation would never have happened on the scale and with the speed that it did so had not the printing press created the potential of making mass quantities of the Bible available in any language. This new availability of Bibles, along with other books, led to an unprecedented explosion in popular demand for literacy skills, and it has been estimated that by 1600 almost one half of the population of London and other major English cities was literate to some degree.

However, it was not only printing which had a decisive effect upon sixteenth-century Christianity, for Greek and Hebrew scholarship also played a vital part. Coming out of the Byzantine world into Italy after 1453 were Greek versions of the New Testament – the language, after all, in which the four Gospels and Letters had been originally written. Then in 1516 the eminent Dutch Catholic humanist scholar and friend of Protestants, Desiderius Erasmus, produced the first printed Greek edition, with a classical Latin translation, of the New Testament, based on a Byzantine Greek version. The original texts of Christianity had now become available in their original tongue. Likewise, Christian scholars began to go to Rabbis to learn

Hebrew, so as to enable them to go back to the earliest sources of the Hebrew Old Testament.

Scholarship on all levels – religious, secular, scientific, medical and philosophical – was advancing faster than ever before. Europe's medieval universities grew, and new universities such as Königsberg (1544) and Ireland's Trinity College, Dublin (1591) came into being. And while the *Trivium* and the *Quadrivium* remained in many ways the structural heart of the curriculum, their contents expanded all the time in the light of Renaissance humanism, as new texts became available and new interpretations were discussed.

Yet printing also brought about a fundamental change in the teaching and study methods of Europe's universities. On the whole, Europe's medieval students imbibed their education not so much by reading as by listening and arguing. Hand-written books, after all, were generally too expensive for 'poor clerks' to own, while library copies were too valuable to be available for borrowing. Instead, a student would often attend the public reading of a specified Latin text, given by a don from a lectern, after which an even more senior don would deliver a lecture upon it. The student would train himself to memorise the spoken word, and would often use memory keys to enable him to do so. Later, he would be called upon to dispute or argue the pros and cons of a set case within the text or lecture. Indeed, the titles of the actors in this academic ritual still survive in our present-day universities, with their *Readers* (who once read out texts) and *Professors* (who professed, or lectured upon them), while graduate students are still required to defend a (now written) *thesis*, followed by a *viva voce*, or 'living voice' cross-examination from the senior academics.

One of the ways in which printing changed the technique of university education was in facilitating the didactic switch away from *listening* to *reading*. With the Aldine and other 'student' editions of the curricular texts, books became sufficiently cheap and plentiful for most students to have access to them. Public readings of texts were no longer required, and both Readers and Professors gradually moved towards a more critical or exploratory style of lecturing, and were less concerned with simple exposition.

The social and geographical basis of education also began to change in the sixteenth century. While the majority of students in Europe's medieval universities had been 'poor clerks' on some kind of scholarship, in a world where the heir to an ancient dukedom could be proud of being an illiterate bred up to fighting, circumstances were changing noticeably by 1550. To be considered a true gentleman by this period, it was becoming necessary to sport culture as well as weapons. An aristocrat who had only birth and brawn was now likely to be seen as old-fashioned and passé; and after 1509, in England the court of King Henry VIII clearly demonstrated these changing social tendencies. Figures like Thomas Howard, the feudal Duke of Norfolk, were perhaps understandably resentful of the social precedence accorded to Thomas Wolsey, the son of an Ipswich butcher, whose scholarship, genius and political sagacity had elevated him to a Cardinalate and Pope's legate, or representative, in England. In King Henry's polished court, highly educated and cultured lads from modest backgrounds – such as Sir Thomas Moore and Thomas Cromwell – rose to the greatest heights, even if such men sometimes lost their heads due to a change in the fickle monarch's whim.

In this Renaissance world, indeed, a *true* aristocrat was increasingly expected to be able to write poetry, sing, discuss the classical authors, display a knowledge of geography and astronomy, and make polite conversation as well as wield a broadsword. To acquire this necessary social polish, wise aristocratic parents took the novel step of sending their sons to university or to London's Inns of Court. Obtaining a degree or legal qualification was not what mattered so much as obtaining a broad education, and books like Baldessare Castiglione's *Book of the Courtier* (1514) spread the fashion for upper-class culture and Italianate social graces. And outrageous as it seemed in certain quarters, even a tiny handful of high-born women were receiving academic educations, albeit from private tutors. Queen Anne Boleyn – King Henry's ill-starred second wife – was a woman of classical learning, with a marked taste for radical Protestant theology. Lady Jane Grey, whose supporters tried to put her on the throne of England in 1553, was also a learned young woman and an intellectual Protestant, while Henry's daughter, Queen Elizabeth I, read and spoke Latin and French fluently from childhood, and was one of the most intellectually brilliant sovereigns in European history.

Yet while the social accessibility of education was percolating upwards, the Protestant Reformation was nevertheless driving wedges between countries whose universities would once have been available to all. For one was now less likely to find an Italian Roman Catholic studying at a newly Protestantised German university such as Heidelberg or Tübingen, in the same way that English parents were becoming fearful of sending their sons to Montpellier or Bologna for fear of their becoming 'infected with popery'. Even so, the radical and cosmopolitan university of Padua, which evaded papal control because of the protection it received from the Serene Republic of Venice, continued to attract northern Protestants because of the excellence of its training in mathematics, science and medicine. The very Protestant Dr William Harvey, the discoverer of the circulation of the blood, underwent his postgraduate medical training in Padua after taking his first degree at Cambridge, before returning to practise in London and making his great discovery.

Lying at the heart of this new scope of, and social attitude to, education were books: newly discovered classical authors now read in the original tongues; recent and contemporary modern poets like Dante, Ariosto or Boccaccio; theological tracts by Luther or Sir Thomas More; alchemical and occult treatises by Paracelsus or Henry Cornelius Agrippa; astronomical books by Robert Recorde; joke books by the physician Dr Andrew Boorde (the original 'Merry

Andrew'); and books of travel, along with printed vernacular Bibles and Prayer Books. The printed word had become the 'internet' of the Renaissance, and as the world drew to its predicted end, ideas seemed to fly abroad with frightful frenzy.

All this change, and the broadening of perspectives, might strike us as 'progressive', and as pointing in the direction of a more recognisably modern world. However, many at the time were in no doubt that the 'new learning' and all of its adjuncts were really no more than significators of Armageddon. Did they not seem to be fulfilling the doom prophecy read in the Biblical Book of *Daniel*, which implied that before the End mankind would run about madly and new signs would appear? And were not new signs and novel ideas increasing at an alarming rate, especially in the sciences of astronomy and geography, both of which had begun their expansion at around the same time as Constantinople fell to the Turks?

Some fifteen years before the fall of Constantinople – though at a time when Ottoman designs upon the city were becoming increasingly clear – the Greek scholar, statesman, and Archbishop of Nicaea, Johannes Bessarion, travelled to Italy to attend the Ecclesiastical Council of Florence of 1438, in very much the same spirit as Manuel Chrysoloras had done more than forty years earlier. Bessarion, who was a native of Trebizond, warmly advocated the union of the long-separated Greek Orthodox and Roman Catholic Churches, although this particular policy made him unpopular back home in Constantinople. He was, however, appointed cardinal by the Pope in 1439, and came to reside in Rome. His beautiful though surprisingly small cardinal's residence, with its still fresh-looking wall paintings, still stands in Rome.

Bessarion's enduring historical significance lay in his patronage of learning and his encouragement of new Greek scholarship in Italy. From a scientific point of view he is remembered as the man who first encouraged in the West the study of Ptolemy – not in Arabic-derived

Latin translations, but in the original Greek. Bessarion wished to take astronomy from its true fountainhead, and at a time when fresh Greek texts were emerging from Constantinople and passing into Italy after 1453, thereby greatly supplementing the Greek manuscripts which he himself had brought to the West, his patronage was to be of the highest importance. Bessarion had himself begun translating the Greek *Almagest* into Latin, although his growing ecclesiastical and diplomatic responsibilities made it impossible for him to complete the project.

It is one of the ironies of history, however, that the Greek-inspired European astronomical renaissance came as an unexpected side-effect of one of Bessarion's diplomatic missions. In 1460 he was sent by the Pope to the Emperor Frederick III in Vienna, as part of an attempt to enlist German and Austrian support for a Catholic crusade aimed at regaining Constantinople for Christendom. The crusade came to nothing, but in Vienna Bessarion met Georg Peurbach – a young mathematician attached to the German Imperial Court – and Peurbach's pupil, Johannes Müller, who is better known to history as Regiomontanus, which was a Latinisation of the name of his native city of Königsberg. Both Peurbach and Regiomontanus – who were respectively in their thirties and twenties when meeting the sixty-year-old Bessarion – were keen to see astronomy improved. In particular, they were aware of errors in existing Ptolemaic astronomy, and of the fact that predictions of eclipses and planetary movements derived from the *Almagest* and the *Alphonsine Tables* were often seriously wrong when checked against the astronomical bodies themselves. This is why astronomy was in need of fundamental overhaul, and why Cardinal Bessarion's move to take the science *ad fontes*, to the original Greek Ptolemy, was deemed so important.

Both Peurbach and Regiomontanus were mathematicians and linguists of the very highest order, and Regiomontanus's brilliance had already impressed Giovanni Bianchini, a senior financial lawyer of Ferrara whose private passion was astronomical calculation. Not only had Regiomontanus solved all the problems set by Bianchini, but he went on to show his middle-aged correspondent how to solve many more.

At this early stage in the European Renaissance, however, people still automatically thought that Ptolemy's astronomy was essentially correct. The task of humanist astronomers was not, therefore, to overthrow the ancients and strike out on new paths of their own, but to rid the wise Greeks of the encrusted errors of mistranslation – from Greek to Syriac, to Arabic, to Latin – that had built up as the *Almagest* had moved through different languages and cultural requirements. They also recognised the need to take Ptolemy's physical constants (such as the precession of the equinoxes) and computational techniques and use them to recalculate the tables of the motions of the Sun, Moon and planets, to root out centuries of possible mistranscription as successive scribes had copied and recopied Ptolemy's tables in this pre-printing age.

To take astronomy back to its pure source, therefore, involved a radically different set of attitudes to the science than that which grew up a couple of centuries later, and which still underpins modern science. Renaissance humanist science was not so much about original research into new realms of knowledge. It was, rather, more like a form of intellectual archaeology, as one made painstaking efforts to obtain a perfect version of an ancient source which was believed to contain, within itself, a key to the truth. Their attitude, therefore, was perhaps closer to that of a modern professional museum curator or a librarian: a person whose first responsibility is to conserve what survives, and then to go on and interpret it.

Bessarion, Peurbach and Regiomontanus realised that what was needed was not only a pure Greek source, but a key to or an abridgement of the *Almagest* which would take the most important sections of this complex treatise and make them more readily available to astronomers and students. Indeed, Peurbach had already commenced this project six years before meeting Bessarion, in 1454, when his *Theoricae Novae Planetarum* (*New Theories of the [Movements of the] Planets*) had attempted such a key based on the familiar Latin version of the *Almagest*.

Under Bessarion's patronage, however, Peurbach began a major abridgement of the Greek *Almagest*, but died suddenly in 1461. Regiomontanus then accompanied Bessarion back to Italy, and took up where his deceased mentor had left off. Regiomontanus's *Epitome* of the *Almagest* became one of the great milestones in Renaissance astronomy, making as it did the most important sections of the *Almagest* available in a definitive Latin translation from the Greek. Regiomontanus dedicated his *Epitome* to Bessarion in 1463, and the work was to play a crucial role in stimulating fresh investigations into planetary astronomy.

Five years before Cardinal Bessarion's death in 1472, Regiomontanus left Italy, and after briefly living in Hungary, settled in the great German commercial city of Nuremburg. But Nuremburg was not only a thriving commercial pivot between Italy and northern Europe; it was also one of Europe's leading centres of technological innovation. The art of printing had already taken firm root in the city, while it was the goldsmith Peter Henlein's 'Nuremburg Eggs' which would soon become one of the world's prototypical spring-driven portable clocks, or watches. It was in Nuremburg that Regiomontanus began to work with the businessman and enthusiastic amateur astronomer, Bernhard Walther.

Very clearly, however, the driving force in Regiomontanus's life was the improvement of theoretical astronomy, based upon the recovered Greek Ptolemy. Being, as he so clearly was, a compulsive calculator whose delight was to solve complex problems in planetary motion across the celestial sphere, Regiomontanus addressed himself to producing the finest astronomical tables that the world had ever known. In 1474, the Nuremburg presses produced both his *Kalendarium* and his *Ephemerides* – volumes of tables that predicted the future movements of the planets, painstakingly computed and elegantly arranged to maximise their ease of handling. Regiomontanus's tables were immediately useful for calendrical and all forms of practical astronomy, and in 1504, Christopher Columbus used the *Kalendarium* to seize a tactical advantage against the natives of Jamaica. During that year, Columbus was on his fourth voyage of exploration to the Americas – and on 29 February he found himself in a tight corner. Realising, however, that Regiomontanus's *Kalendarium* predicted a lunar eclipse for that night, he was able to convince the local population of his special power in using his prayers to restore the Moon's light.

Regiomontanus's work in recovering Greek astronomy was vital; but one might argue that he made an even more important contribution, for he was the first person to realise the significance of printing for the new astronomy, not just in terms of the typographical fixity of printed texts, but of astronomical tables in particular. Indeed, he was quick to realise that while an error in a manually transcribed table could cause a great deal of mischief, a *printed* error could produce mischief across the length and breadth of scientific Europe. He therefore urged the production of special fonts of type designed for astronomical printing, as well as the need for scrupulous proof-reading before the *Kalendarium*, *Ephemerides* and similar works went to press. The printed book opened up possibilities to late fifteenth-century Europe of so profound a character as would scarcely be transcended until the invention of computers and the introduction of the Internet five hundred years later.

In addition to having definitive Greek texts and superior tables for astronomical calculation, Regiomontanus realised that if the celestial movements were to be properly understood then it was also necessary to observe them with accurate mathematical instruments, as a

way of comparing the predicted and the actual positions of the planets. From as early as 1462, he was making direct observations of the
sky, and one of his favouite instruments was Ptolemy's Rulers, or
triquetum. This instrument (which was referred to briefly in Chapter
5) made use of the natural geometrical properties of triangles – a
department of geometry which especially fascinated Regiomontanus.
But how did these Rulers work?

If one takes three wooden rods, of exactly the same length, and
arranges them into a triangle, then each of the three internal angles of
the triangle will contain 60 degrees. These three rods can also be seen
as relating to a circle. Two of the rods form radii to that circle, and

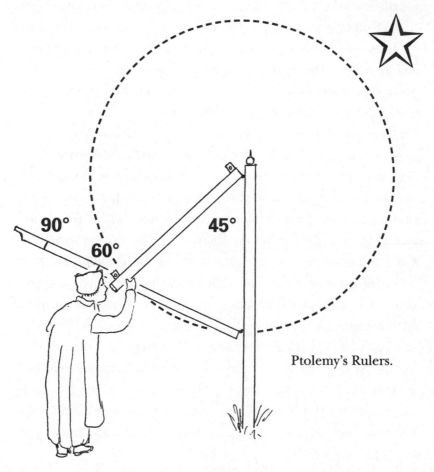

Ptolemy's Rulers.

are pivoted together at the circle's centre, while the other rod connects the two radial rods to form a *chord* to the circle. If the two radial rods always remain the same length, and are pivoted together at the centre of the circle, then when one widens the angle at the centre beyond 60 degrees, the chord rod will also become longer, and the opposite will happen if the angle is taken below 60 degrees. If a set of equidistant divisions have been drawn upon the chord rod, it can be used in conjunction with a precalculated table to yield the exact amount for the central angle, for as the two radial rods are moved apart through every angle from 0 to 90 degrees, their own unchanging length will always have an exact proportional relationship with the length read off on the chord rod for any given angle. If a pair of sights is attached to one of these radial rods, so that one can look at a star or planet, it becomes possible to use the rods to measure angles of elevation above the horizon (as seen in the drawing overleaf).

These great wooden Rulers had two enormous advantages, considering the technology in 1460: they were simple to make, and they were capable of remarkable accuracy. All that one needed was a good carpenter and metal worker to plane the rods straight, make their brass pivots, and set them up in the vertical with a plumb line. And as no complex circular divisions were needed, it was relatively easy to lay off the equidistant straight divisions with a ruler and a pair of dividers. European astronomers now had an instrument that was much more accurate than the long-familiar astrolabe, and with it they could monitor the movements of the planets and compare them with their predicted places as obtained from tables derived from Ptolemy.

Peurbach and Regiomontanus had become the first modern Europeans to recognise the need to actually observe the heavens with accurate instruments if one was to truly take astronomy back to its fountainhead. But it was Regiomontanus' Nuremburg friend, Bernhard Walther, who would be the first north European to begin the practice of making regular and sustained astronomical observations from one place. Between 1475 and his death in 1504, Walther observed the daily altitude of the Sun, the position of the Moon, and the motions of the planets against the stars. He used a set of Ptolemy's Rulers, a Radius Astronomicus (a variant version of the Rulers) with rods over eight feet long, and an armillary sphere that was similar to Ptolemy's. For twenty-nine years, and for long after Regiomontanus's death in 1476, Walther built up that arsenal of primary physical data against which theories could be tested, and new advances made. His observations would later be used by Copernicus and many subsequent generations of astronomers, and would lay the foundations of modern European observational astronomy (as we shall see in Chapter 10).

Just as the new developments in astronomy could be interpreted as fulfilling the *Daniel* and *Revelation* prophecies that, before the end of the world, unfamiliar signs would appear in the heavens, so contemporary geographical discoveries seemed a perfect fulfilment of that restlessness and searching which would precede Armageddon. After 1460, new geographical discoveries came thick and fast, culminating sixty years later in the first circumnavigation of the globe, by Ferdinand Magellan's expedition. On the other hand, medieval geography was nowhere near as backward as is popularly believed (as will no doubt have gradually become apparent throughout the course of this book). For example, after 1250 – by which time Aristotle's *De Caelo* and Ptolemy's *Almagest* were being assimilated into mainstream European culture – no educated person believed that the Earth was anything other than a sphere. They might, however, have been uncertain exactly how big it was, and how much of its surface was covered with land and how much with water.

Generally speaking, however, people believed that the Asiatic land mass occupied a much bigger percentage of the Earth's surface than in reality – a belief largely guided by Ptolemy's other great treatise, *Geographia* (*The Geography*), which only came to be known in

Europe in a Latin translation from a Greek original in 1406. And was there a vast, unknown continent in the southern hemisphere – the legendary *Terra Incognita Australis* – that perhaps joined on to Africa? Did Africa contain an immense lake that fed the waters of the river Nile? And was Ethiopia the domain of Prester John, the great Christian king of Africa? And of course, no-one in Europe knew that the American continent and the Pacific Ocean existed, so that it did not seem unreasonable to assume that the African shores of the Indian Ocean – known from classical and Arabic sources – formed the western extremity of that same ocean which far to the east lapped the rocks of Cornwall.

What, therefore, gave rise to the popular belief that medieval scholars thought the Earth was flat? Well, in addition to the popular tenacity of 'dark ages' myths, and the writings of the non-scientific early theologian Lactantius, one set of surviving artefacts could, when read out of context, undoubtedly conduce to this belief. These were the *Mappae Mundi*, or maps of the world, one of the finest surviving specimens of the genre being the thirteenth-century Hereford Cathedral map. Mappae Mundi were large wall-maps, usually painted on a great skin of animal vellum. They showed God reigning over a flat, dinner-plate-like world, in which Jerusalem was always at the centre, and in which the continents and cities of the known and imagined world were spread out in a seemingly fanciful distortion towards a peripheral and all-encompassing ocean.

Yet these Mappae Mundi were not intended to be scientific so much as religious geographies, showing God's governance of Creation, at the heart of which lay Jerusalem, the location of Christ's crucifixion and resurrection. Mappae Mundi were spiritual exemplars and guides to heaven, and were no more intended to depict the *natural* world than is a map of the London Underground – while capable of guiding you from Epping to Brixton via Oxford Circus – intended to be a naturalistic map of London.

Indeed, if one wants to obtain a real sense of how medieval people envisaged the face of the Earth – or at least those parts of it with which they had direct experience – one should look at the *Portolano* charts, used by navigators from 1300 onwards. Portolano charts depict the coastlines of Europe and the Mediterranean, from Ireland and Iceland to Sicily and Byzantium. They contain sailing direction lines based on the thirty-two points of the compass, and are often remarkably accurate in their depiction of bays, headlands and distances. From a good Portolano – each one of which was projected and drawn as an individual scientific work of art on a large piece of vellum – a fourteenth-century navigator could lay off his course from, let us say, the coast of Sligo in Ireland, across the Bay of Biscay to Lisbon, Barcelona, Marseilles and Palermo, through the Greek islands, and on to Constantinople. Venice was one of the foremost centres of their manufacture, and we know of individual cartographical draughtsmen from as early as the 1320s, such as the Venetian Petrus Vesconte, who were making them. While it is true that these Portolano charts cover only a relatively small part of the Earth's surface, from the Canary Islands in the west to the Black Sea in the east, they nonetheless indicate the existence of a naturalistic approach to geography in the middle ages that works alongside and serves a different purpose from that of the spiritually symbolic Mappae Mundi.

With the rediscovery of Ptolemy's *Geographia* in the West and its translation into Latin, however, crucial new technological innovations took place in cartography that would be of the greatest importance to the post-1460 navigators. One of the most significant of these was Ptolemy's use of latitude and longitude, and another was his technique of using timed observations of eclipses of the Sun and Moon to establish the precise longitudinal differences between places on the Earth's surface. While pre-Ptolemaic European astronomers were familiar with the use of the quadrant and the astrolabe to find their *latitude* from measured observations of the height of the Pole Star,

Ptolemy's method of using the *longitude* supplied in their basic form by the early fifteenth century the techniques both of navigation and of map-drawing which we still use today.

Yet all of these cartographic and navigational techniques, including those recently obtained from Ptolemy's *Geographia*, were still seen as pertaining to the old world of Europe and the Mediterranean. What factors were at work, therefore, to make Europe burst the geographical bounds established by the classical geographers, and, in these seeming 'last days', sail out to discover the rest of the world?

When looking at the situation of Europe as it was perceived by fifteenth-century people, one sees a cultural see-saw motion in action; for as the rising power of Islam was threatening to bring Armageddon from the east, the peoples of the West – especially those of the Iberian Peninsula – were looking towards the sea for their worldly hopes. Long before Christian Constantinople finally fell in 1453, the rise of Ottoman power had become increasingly ominous to Europeans, and one of its first effects had been economic. While we may think of economic warfare as a being modern, it is in reality as old as trade itself, and from the late fourteenth century onwards, the Near-Eastern Muslim Princes who controlled the Oriental spice trade were restricting supplies and forcing up the prices of these vital culinary commodities.

It was this initial pressure which, in 1418, led Prince Henry of Portugal – known to history as Henry the Navigator – to send his captains on exploratory voyages into the Atlantic and down the coast of Africa. If one could somehow find a sea route to the East, one could perhaps find the source of the spice trade, cut out the Muslim middle-men, and bring the spices back wholesale into Europe.

Towards this end, Prince Henry further established Europe's first school of navigation, cartography and practical astronomy at Cape Sagres, Portugal, to train his explorers in a scientific manner. Not only were these men made familiar with the use of Portolano charts, the magnetic compass, the astrolabe and other astronomical instruments; they were also sailing in ships of a new design. Medieval Europe's commercial workhorse of the sea – as depicted in numerous city seals and official documents of the period – was the *Cog* – a single-masted, round-hulled vessel. By the 1450s, however, the Caravel and the Carrack began to emerge. These three-masted vessels had longer, narrower hulls, and rudders of a more efficient design, and were much more seaworthy than were cogs. Without these rugged vessels, which were capable of holding their own in stormy seas for months at a time, fifteenth-century Europe could never have set out upon its great oceanic adventures, for it would have lacked the transport technology to do so.

The search for the source of the spice trade, therefore, was a powerful motivating force behind the Iberian Atlantic explorations; although there was another, which became especially urgent after the fall of Constantinople in 1453. This was the search for Prester John, who according to legend was a Christian Priest–King who ruled a vast African Empire somewhere in the region of Ethiopia – although some said that his kingdom was in central Asia! While no such kingdom ever existed, the legends nonetheless had some basis in reality, for Coptic and Nestorian Christian communities still exist in Egypt, Ethiopia and Persia. What Prince Henry and, after 1460, his royal successors John and Manoel, hoped was for Roman Catholic Christendom to form an alliance with Prester John, to launch a joint crusade against the Turks and to relieve western Europe from its perceived threat. This is why the Iberian Oceanic adventure was often described as being in search of 'Christians and spices'. Prester John was, understandably, never found, and (as mentioned above) mainland Europe was never invaded by the Turks; but the Oceanic adventure became one of the most potent and far-reaching products of the scientific renaissance.

During Prince Henry's lifetime (before 1460) the graduates of the Sagres Academy had discovered the Cape Verde and Azores groups of Atlantic islands, as well as Madeira. The Portuguese then began to edge down the coast of Africa, where they made a succession of discoveries which openly contradicted the classical geographical writers. For one thing, they found that when human beings crossed the equator of an indisputably spherical Earth, their blood neither boiled in their veins, nor did they go mad. Nor were their ships swallowed up by monsters. And, as Bartholomew Diaz discovered in 1487, Africa was *not* joined to the supposed Antarctic land mass *Terra Incognita Australis*, but rather, the Atlantic and Indian Oceans came together to form one body of water at 35 degrees south of the equator. And as Vasco da Gama discovered in 1498, one could sail from Portugal, round Africa, and on to India and Ceylon (Sri Lanka) – to the supposed sources of the spice trade. From 1500 onwards, therefore, Europe became aware of the enormous wealth of the 'East Indies' that lay within its commercial grasp.

At the same time that Portugal was finding a route to the East by sailing down the Atlantic, so her rival Iberian power, Spain, was growing in strength. Spain's formidable royal partnership, Ferdinand and Isabella, had, in their marriage in 1469, joined the two great dynasties of Castile and Aragon to unite Spain under one rule. They also completed the centuries of Re-Conquista by taking the last Islamic princedom in the Iberian peninsula, Granada, in 1492, so that, ironically, as Muslim power grew on Europe's eastern flanks, it was extinguished in the West. And carefully eyeing their Portuguese neighbours, Ferdinand and Isabella also formed designs upon the Atlantic.

Spain's Atlantic die was eventually cast when, in 1492, a Genoese sea captain named Christopher Columbus approached Ferdinand and Isabella to ask for backing for a voyage to find China by sailing due west across the Atlantic. Ferdinand and Isabella were by no means the first European monarchs whom Columbus had approached, as he had already received refusals from France and from King Henry Tudor in England. Columbus's scheme did, after all, seem replete with risks, for while the Portuguese had shown that coastal west Africa could be navigated, no-one knew what terrors lay between Europe and the Orient, which was believed to lie due west. However, one thing which no-one believed lay across the Atlantic, and yet the myth of which somehow became firmly entrenched in the modern American public education system, was the edge of the world over which the brave Columbus was in danger of sailing! For as we have already seen, no educated person in 1492 believed the world to be anything other than a sphere.

The questions upon which Columbus's proposed voyage turned, rather, were the size of the Asiatic land mass, the ratios of land and sea on the Earth's surface, and the relative proximity of China to Spain across the Atlantic. Columbus's proposed route to Asia hinged upon a number of postulates: the terrestrial diameter (and hence circumference) was thought to be about 20% smaller than we now know it to be. Asia, moreover, was believed to occupy much more of the Earth's surface and to come much further east – towards Spain – than would be discovered to be the case. And Columbus pinned some of his faith on the above-mentioned passage in the apocryphal Biblical Book of *Esdras* II (6:42), which said that God had covered only one seventh of the Earth's surface with water, thereby making the Atlantic a lake (relatively speaking). And while learned theologians cast doubts on the canonical status of *Esdras*'s pronouncement, Columbus was quick to point out that St Augustine himself had taken seriously the prophecies of the Book of *Esdras*.

Errors in classical geography and disputes over the canonical status of Biblical passages, therefore, form an essential background against which to understand Columbus's voyage. Knowing what we now know about the size of the Earth, Columbus, and the leaky *Pinta*, *Nina* and *Santa Maria* which Ferdinand and Isabella decided to risk on his map-cap venture, should have come to grief in the uncharted wastes of a 15,000-mile-wide ocean had not the American continent been in the way!

On 3 August 1492, Columbus sailed from Spain with his three small ships and 120 men, an astrolabe, and other instruments, and after taking on fresh provisions off the Canary Islands, reached San Salvador (Watling Island) in the Bahamas on 12 October. He then went on to discover Haiti and Cuba, and was back in Palos, Spain, by mid-March 1493. He was to make three further voyages to the New World – as his newly discovered Caribbean islands and the coasts of Mexico came to be known – before his death in 1506, although ironically he went to his grave firmly believing that what he had discovered were the islands off the coast of China, and not a new continent.

It was Columbus's lieutenants and successors who realised that what lay 3,000 miles west of Spain were not the islands off China, but the hitherto unknown continent of America. And while it came to be realised by Scandinavian scholars centuries later that North America could well have been Vinland, or the land of grapevines, discovered by the Norwegian Leif Ericson around AD 1000 and described in the *Vinland Saga*, this information was not known to the people of Europe beyond Scandinavia in the 1490s, and played no part in Renaissance Europe's burgeoning awareness of the wider world.

Nor are scholars by any means agreed as to the derivation of the name 'America' for this New World. While it is generally believed to be derived from the Christian name of Amerigo Vespucci – the Florentine gentleman and Spanish agent for the Medici, who sailed with Columbus in 1497 – we also know that the wealthy Bristol merchant venturer, Richard Ameryck, was helping to direct voyages undertaken by the Genoese Cabot family to and beyond the cod-rich Grand Banks of Newfoundland before 1500. Indeed, cod were big money in late medieval Europe, for when every Christian was required to eat fish on Fridays, knowledge of a seemingly inexhaustible supply of fish – which could be dried and salted for preservation and export – was, for the Bristol merchants, worth a high-risk capital investment.

One of the remarkable things about the post-1487 geographical discoveries is the speed with which they were communicated to the wider world. Indeed, this is all the more remarkable when one remembers how potentially lucrative these new trade routes could be to their discoverers. The world map of Giovanni Matteo Contarini of 1506, for instance, is the first printed and published chart – a north polar projection – to show Columbus's discoveries. And while at this early date America is still being depicted as a part of China, Contarini's map of Africa is astonishingly accurate. And then, in 1500, some manuscript maps showing Columbus's discoveries were captured from an Italian vessel by Turkish pirates in the western Mediterranean. These manuscripts were taken to Constantinople, where the information which they contained was incorporated into the 1513 world map of the Turkish admiral Piri Reis.

The Piri Reis map – now preserved in the Topkapi Palace Museum, Istanbul (Constantinople) – is a beautiful conflation of old and new geography, drawn in the form of a Turkish Portolano chart, complete with compass bearings. It consists of the Atlantic fragment of what was probably once a Turkish world chart. Spain and western Africa are well depicted in the east, and in the west, Brazil and the northern Caribbean are clearly drawn; but further north, the geography becomes confused. The island of Hispaniola, for instance, is called Cipangu, the medieval European name for Japan, while other Caribbean islands were also mistakenly given oriental provenances. All of this is quite understandable, however, since Columbus himself believed the Caribbean islands to be part of a group that stood just off the Chinese land mass. Indeed, scholars now believe that the Piri Reis map contains, in second- or third-hand transcriptions, some of the earliest pre-1498 cartographic depictions of the Americas taken directly from captured copies of Columbus's own manuscript survey charts. Continuing down south of Brazil, instead of tapering to its as yet undiscovered point, South

America suddenly veers south-east, to form part of *Terra Incognita Australis*, where it presumably becomes attached to the Asiatic land mass. The Atlantic and Pacific Oceans are therefore conflated on the Piri Reis Map, and the Chinese and Japanese seas lie somewhere in the region of present-day New York and Boston. This was, of course, the world as Columbus believed it to be.

By the time of the return of Ferdinand Magellan's expedition to circumnavigate the globe in 1522, however, the world as we now know it was essentially there. America was recognised to be a vast continent in its own right, incorporating the Caribbean; and the Atlantic, Pacific, Indian and other oceanic masses were in their correct places, and thereby covered most of the surface of the globe with salt water. It is true that the discovery of Australia proper lay more than two centuries in the future, and a colossal amount of detail had still to be inserted into the *totus terrarum orbis*, or complete map of the world; yet in no more than one average human lifetime, European people's knowledge of the surface of the planet on which they lived had changed beyond recognition.

But what did all of this new knowledge really mean? Were the new astronomy and geography simply significators of the speed and novelty which would precede Armageddon? Or could it really be that modern humanity – which everyone admitted was the shrunken runt-end of the divine creation, nowhere near as long-lived as Adam or Methusalah (who had walked the newly created Earth), nor as wise as Plato and Aristotle – had really discovered something of enduring importance?

Indeed, if one wished to take a fatalistic view of the new scientific discoveries, it was very easy to do so, for geography in particular seemed redolent of impudence. Instead of respectfully following the logical and philosophical rules of their classical mentors, were not the early explorers, with their new-fangled ships and charts, behaving like boisterous children plunging headlong into folly? Did God really

want Europeans to sail the Pacific, or was this no more than a signifi-
cator of that Pride which would come before the Fall?

On the other hand, one could see something wonderfully positive in the whole business. Had not Bessarion's and Regiomontanus's
astronomical discoveries been made with the explicit intention of
obtaining a better understanding of God's creation? And had not the
explorers earnestly desired to make contact with fellow Christians in
Africa and Asia, as well as to trade in spices? As these discoveries in
astronomy and seamanship, moreover, had been made not by
philosophising but by the use of physical objects and instruments,
could it not also be possible that the human senses of sight, hearing,
touch and smell were equally valid modes of discovering the truth, as
well as simply following the laws of logic and reading the oldest texts?
Geography is, after all, a supremely empirical, hands-on science, for
no philosopher in his study could have mapped the Caribbean in the
way that Columbus mapped it with his ships.

But perhaps most challenging of all was the resulting reassess-
ment of the relationship between the 'Ancients' and the 'Moderns'.
For if modern man could uncover new knowledge in astronomy,
geography, anatomy, optics, and so on, all of which glorified their
Creator God, could the men and women of these 'last days' really be
history's runts? Or could they – dare one suggest – be even wiser and
become more learned than the people of antiquity had ever imagined?

It was one of the paradoxes of the Renaissance, therefore, that
while on one level – as seen in the fall of Constantinople, the appearance of new diseases such as syphilis, and the questionable morality of
certain Popes – the Apocalypse seemed genuinely at hand, on another
level new creative forces were breaking loose. From these ashes at the
end of time sprang not only a Protestant Reformation, but also
dynamic new moves within the Catholic Church itself to put its own
house in order, combined with a rediscovery of the fountainhead of
classical learning and a new appreciation of rational truth. But perhaps

most enduring of all, as the Byzantine East was lost, Europe looked increasingly to the West and found a new avenue for its creative energies upon the broad waters of the world.

T E N

The astronomical revolution

Sixteenth-century European astronomy was dominated by two considerations – one of them essentially intellectual and theoretical, and the other spiritual and practical. The first related to the continuing yet still inadequately explained problem of planetary motion. The second was concerned with the calendar which, in spite of Sosogenes' leap years and Bede's improved techniques for finding the date of Easter, continued to run into error at the rate of one day in just over a century because of the precession of the equinoxes. Yet both sets of problems were intimately related, and were handled by and large by the same men, most of whom were either clergy or else enjoyed the patronage of the Roman Catholic or the new Reformed Churches.

It might reasonably have been expected, however, that by the early decades of the sixteenth century – some fifty years after the Greek-based humanist astronomy of Cardinal Bessarion, Peurbach and Regiomontanus had had time to percolate to the very heart of learned Europe – the problems of planetary motion should have been solved. Good Greek texts, superior printed astronomical tables distributed across the universities of Europe, better techniques of mathematical computation, and a more widespread use of instruments based upon those described in Ptolemy, should have finally reduced the heavens to predictable and orderly understanding. Yet this was found not to be the case, and the more astronomers began to test the predictions extracted from these undoubtedly superior tables against observations of the sky, the more errors they found.

All of these astronomers were, and would be for the next century, still working within those classical Greek assumptions about the heavens that we have already encountered: namely, that the astronomical bodies move at perfectly uniform speeds, in perfectly circular orbits, around the immovable Earth. However, none of these theoretical assumptions actually exist in nature – in reality, the planets move at variable speeds, in *elliptical* orbits, around an Earth which itself rotates around the Sun – and it is therefore hardly surprising that these predictions never worked.

The man who was destined to begin the resolution of this deadlock was a Polish cathedral dignitary of conservative habits who held Ptolemy and the ancients in the highest possible regard. It is not without irony that the man who would ultimately open the door to the eventual abandonment of classical Greek celestial mechanics saw his academic vocation as the reinterpretation and final consolidation of the Greek tradition.

Nicholas Copernicus (1473–1543) was born at Torun, Poland, into a well-to-do family, the son of Mikolaj (Nicholas) and Barbara Kopernik, although he followed the custom of Latinising his name to the form with which the world would become familiar. His father and his maternal uncle Lukasc Watzenrode had both been active in Polish resistance against the depredations of the Teutonic Knights, and as his parents died while he was still only a boy, young Nicholas entered the household of his powerful uncle. Lukasc Watzenrode was the Roman Catholic Bishop of Ermeland, and was by definition a Polish feudal magnate of great influence. Very clearly, Bishop Lukasc recognised his young nephew's talents, and clearly intended to train him for a senior post in the Church, which in those times also meant being an important civil administrator.

At the age of eighteen, in 1491, Nicholas was sent to the Polish University of Cracow, where he would have become familiar with the Latin-based *Trivium* and *Quadrivium* curriculum, as well as with aspects of the new Greek learning. Then, in 1496 uncle Lukasc sent him to Italy, to study for what we would now call a postgraduate degree in law at the great University of Bologna.

It was probably in Italy that the young law student started to take a serious interest in astronomy, for on 9 March 1497 he observed a lunar occultation of the star Aldebaran, in the constellation of Taurus. Such occultations (the Moon's passage in front of stars) were important to astronomers who were trying to determine a predictable orbit for the Moon, for while the lunar orbit is complex and wandering, the stars never change their positions, so that one can use the place of the occulted star to derive an exact timed and dated position for the Moon. We also know that the 27-year-old law student was in Rome on 6 November 1500, for he recorded observing a lunar eclipse there on that date. Lunar eclipses are also important for understanding the Sun's and Moon's orbits, for when such an eclipse takes place the astronomer knows that the two bodies are exactly 180 degrees apart, with the Earth on the straight line between them.

In 1501 Nicholas was back in Poland, for even though he was never to become a full priest, his uncle Lukasc had already raised him to the dignity of a Canonry of Frombork (Frauenberg) Cathedral. The Cathedral's Chapter, or governing body, of which Canon Nicholas was a member, then released him to return to Padua to study medicine, and he finally returned to Poland for good in 1503.

During his student years, Copernicus imbibed the Greek learning from its most active source: Italy. He bought books, and it says something about his already considerable standing as an astronomer that he was even consulted by the Pope on matters of calendar reform. The real purpose of this educational investment, however, was to make him especially useful as a humanist scholar to the life and administration of his native Poland, and in particular, Frombork Cathedral Chapter. One notices, however, that his formal training was in law and medicine; though being literate in both Latin and Greek, and with an obvious head for mathematics, he had little difficulty in acquiring his higher astronomical knowledge from private study and from extracurricular contacts.

We know very little of Copernicus's astronomical activities after his return to Poland in 1503, though there is record from ten years later – 31 March 1513 – of his purchase of 800 building stones with which to build an observatory. When finished, this housed a set of Ptolemy's Rulers of 4 cubits, or just over 5 feet radius, and a quadrant, both of which were used for measuring the heights of astronomical bodies above the horizon; and also an astrolabe, which he no doubt employed to calculate spherical triangles, or the mutual angular relationships of planets as seen against the curved imaginary sphere of the sky.

It is, indeed, quite possible to reconstruct what Copernicus was doing at this time. From his knowledge of Ptolemy – as refined through the *Epitome*, *Ephemerides* and *Kalendarium* of Regiomontanus, and the medieval *Alphonsine Tables* of which he owned a copy, and from his own observations of the heavens – he was trying to establish a consistent model of the heavens in which theoretical prediction and factual observation matched exactly. But he failed, primarily because (as we have seen previously) the *real* Universe is not built as the Greeks believed it to be.

Also around 1513, Copernicus began to peruse the works of other Greek astronomers and philosophers for ideas, and, as he tells us in 1543, he found that several classical figures – including Philolaus, Heracleitus and Hicetas – had all discussed the possible motion of the Earth. And while we are not certain about Copernicus's familiarity with the writings of Aristarchus at this time, it soon came to be realised that Aristarchus of Samos had also proposed a heliocentric system around 270 BC. By 1514, however, Copernicus had already produced a carefully devised Sun-centred (heliocentric) arrangement for the Universe, as recorded in his *Commentariolus* of that year. This work, however, was not intended for publication, although it might have been seen and perhaps discussed by a select number of Copernicus's colleagues.

Yet what scientific advantages did a heliocentric theory have over the geocentric theory of Ptolemy and Aristotle? One of its greatest profits was its ability to provide a relatively simple solution to the retrograde loops of the planets, which had plagued astronomers since the sixth century BC, and which, over the centuries, had led to the invention of the complex arrangement of epicycles and eccentric circles to explain them.

If one thinks of the Earth as rotating around the Sun in one year, Mars in two years, Jupiter in twelve, and Saturn in almost thirty, then it is clear that, from our faster-moving planet Earth, we will catch up with and then streak ahead of each planet two, twelve, and thirty times respectively, as Mars, Jupiter and Saturn complete their own full orbits around the Sun. As we seem to catch up with a slower-moving planet, it will create the impression of its moving *backwards* when seen against the background of unchanging stars. Yet when we see that same planet from across the solar system – as the Earth seems to 'corner' in its orbit – the resulting line-of-sight effect will cause the illusion that the planet is suddenly *accelerating*. And between its apparent bursts forwards and backwards, it will seem to be stationary.

Copernicus's explanation was elegant and simple, and he claimed that one of the factors which had led him towards the heliocentric theory was the apparent Earth-year basis for the planetary loops. However, his theory was far from all-embracing in its explanatory power, for he still worked upon the classical assumption that the planetary orbits must be perfectly circular, and the planets' speeds within them perfectly uniform. Over the next three decades, therefore, he tried configurations of epicycles and eccentric circles in his attempts to fine-tune his system.

In addition to his purely Ptolemaic and Regiomontanian mechanisms, however, it would be interesting to know how well acquainted Copernicus was with the achievements of the Maragha School and the Damascan Arab astronomers of the thirteenth century; for in his use of eccentric circles and regularly sliding orbital radii to make circular motions produce non-circular effects, he was employing geometrical constructions that were virtually identical to those proposed by Nasir Al-Din Al-Tusi and his pupils three centuries earlier. However, there exists no firm proof that Copernicus was aware of this brilliant school of Arab celestial mechanicians (mentioned in Chapter 6).

In the midst of his clerical administrative duties in Frombork, however, Copernicus made regular astronomical observations – at least up to about 1530 – as part of that data accumulation and theory-testing which were to give weight to his heliocentric hypothesis. He was aware, though, that his theory laboured under a massive disadvantage: neither astronomical observation nor general human experience suggested that the Earth moved around the Sun. If it did move, moreover, this must surely undermine the entire physics of Aristotle and the doctrine of four purely terrestrial elements, which depended on the premise that the Earth was immobile.

This lack of both physical and mathematical proof of the Earth's motion was the reason why Copernicus was so cautious about publishing. He feared academic ridicule, throughout the universities of Europe, for proposing a theory which, in return for explaining the planetary loops, undermined the whole of common-sense physics. But what he did *not* fear was ecclesiastical persecution, for not only was he a highly respected Cathedral dignitary and a sincerely devout

Roman Catholic, but as he was all too well aware, the Church had no especial dogmatic opinions on science in 1540. His concerns about credibility, therefore, were related to the purely academic and not to the ecclesiastical world.

Then, in 1539, a young Protestant mathematician, Georg Joachim Rheticus, of Wittenburg University, visited Copernicus and was allowed to read the manuscript of the 66-year-old Canon's *De Revolutionibus Orbium Coelestium* (*On the Revolutions of the Celestial Spheres*). Rheticus was deeply impressed, and urged Copernicus to publish, although he did not have the time to see the book into print himself. In 1540, however, he wrote an open letter to Johann Schoener, of Nuremburg, and this was published in Dantzig (Gdansk) under the title *Narratio Prima*, or 'first communication' of Copernicus's work.

The publication of *De Revolutionibus*, however, fell to another Protestant, Andreas Osiander, from whose hands the book was given to the world in 1543 – but not before Osiander had added an unauthorised Preface which said what, almost certainly, Copernicus himself had not believed. Osiander's Preface claimed that the heliocentric theory was not intended to be a factual description of the structure of the heavens, but a 'saving of the phenomena' – a purely explanatory device, or a hypothesis which made it easier to calculate tables.

What Copernicus himself thought of Osiander's highly contentious and subsequently litigated Preface is not recorded. *De Revolutionibus* was printed in Nuremburg during the winter and spring of 1543, and Copernicus, residing in Frombork, no doubt never saw the proof sheets. Besides, in December 1542 Copernicus suffered a major and debilitating stroke, and according to legend his first sighting of his *magnum opus* was on his death bed, on 24 May 1543. And yet what is particularly remarkable about the publication history of Copernicus's ideas is its cosmopolitan character, at least within the Germanic cultural sphere: written by a Roman Catholic in

Frombork, praised in Lutheran Wittenburg, first officially noted in Dantzig, and finally published in Protestant Nuremberg.

What Osiander's Preface brings to mind, however, is that previously mentioned creative tension or paradoxical view of knowledge first encountered by the Greeks. For is science about the actual description of nature as it is, or as it faithfully appears to our senses? Or is it about 'models' or 'hypotheses' that provide adequate mathematical descriptions of phenomena such as the motions of the planets, without being able to say for certain what *really* rotates around what? Of course, the true aspiration of science is to combine them both, and to devise precise mathematical explanations of the phenomena as they really exist. For almost two centuries after Copernicus's death, however, it was still impossible to prove by direct observation that the Earth really *did* move around the Sun, although long before James Bradley's decisive proof – which derived from his discovery of the aberration of light in 1728 – a growing accumulation of supplementary arguments had convinced Europe's scientists beyond doubt that this was indeed the case.

Not only had the young Copernicus advised an early sixteenth-century Papal Commission about important parameters necessary in the reform of the now badly erroneous Julian Calendar, but eight decades later this same Church would use data taken from *De Revolutionibus* and incorporate it into the eventually 'reformed' calendar of 1582. For by the time that Pope Gregory XIII proclaimed the reformed, or Gregorian, calendar in that year, the Church's finest mathematical intellects had been collecting data and refining formulae for several centuries, and since the thirteenth century in particular.

Yet fundamental to any *permanent* reform of the calendar, or one that would not require yet another overhaul in the centuries to come, was the need to know the length of the year with critical accuracy – and this in itself demanded a feat of practical observation of the highest order. To obtain this figure one had to know the *exact* highest and lowest points to which the Sun ascends and descends at the summer and winter solstices, and its intermediary spring and autumn equinoxes, and in 1550 no quadrant or set of Ptolemy's Rulers in Europe was accurate enough to determine these figures.

Then, in 1574, Egnatio Danti – a Dominican priest and mathematician at the Florentine Court of Cosimo Medici – set up a vertical dial on the façade of the Church of Santa Maria Novella in Florence, in the hope that from its readings, over the course of the seasons, he might extract the necessary value for the length of the 'tropical year'. On moving to the Dominican House in Bologna in 1575, Danti obtained permission to set up a 'meridiana' instrument – a massive sundial – in the great church of San Petronio, Bologna.

Egnatio Danti's *meridiana* consisted of a small hole punched through the south masonry wall of the San Petronio church, some 65 feet above the pavement, so that at noon each day a shaft of sunlight burst through the hole, to fall upon a carefully surveyed north–south line laid out on the marble floor. With this massive projection distance it was possible (at least in theory) to obtain extremely accurate values for the features of the solar orbit, although in practice, all sorts of factors – such as slight irregularities in the floor level on which the solar image fell – could spoil the results.

Even so, the construction of these *meridianae* in churches in Bologna, Rome, Florence and elsewhere provide an indication of how important calendrical astronomy had become to the Church. Indeed, it would be difficult to find better examples of the Renaissance Church's active concern with precision astronomy than in its development and perfection of the Gregorian Calendar after 1582, combined with the readiness of Popes and Bishops to allow the sacred fabric of their great churches to be used as astronomical observatories by scientific priests. For long after the introduction of the new calendar, successive generations of mainly Italian astronomers would use these great ecclesiastical *meridianae* as a way of refining their values for the solar motion and maintaining the calendar's correspondence with the heavens.

In addition to matters surrounding the calendar, Europe's astronomers responded with alacrity to Copernicus's *De Revolutionibus*, and the book became so sought after that a second edition was printed in Basle in 1566. Copernicus's book provoked debate because, on the one hand, it genuinely facilitated the calculation of planetary tables, while on the other, the idea of an Earth that not only rotated around the Sun, but also spun upon its axis once per day, seemed an affront both to common sense and to Aristotle's science. Even so, Copernicus won friends across Europe – not least in England, where the Welsh physician Robert Recorde announced his support in the 1550s, and Thomas Digges published, in 1576, an English synopsis of the Copernican theory in his *A Perfit Description of the Caelestiall Orbes according to... Copernicus*, complete with an illustration of the Sun-centred solar system. Yet Digges even goes beyond Copernicus in suggesting that, instead of the stars all being at a fixed distance from the Sun, the Universe could well be infinite, for 'This orbe of starres fixed infinitely up extendeth itself in an altitude sphericallye, and therefore immovable: the pallace of felicitye farre excelling our sonne both in quantitye and qualitye the very court of caellestiall angelles devoyd of greefe and replenished with perfit endlesse joye the habitacle of the elect.'

It is clear that Digges drew deep inspiration from the new sciences, and from Copernicanism in particular, arguing that the infinite Universe of Sun-like stars rose up to heaven itself to form an elegant fusion of the scientific and the theological. Yet it was a development of this infinite Universe idea, combined with elements borrowed from Hermitic occult philosophy, and a general pantheism in which God became depersonalised and Christ ceased to be divine, which in 1600 was to cost Giodarno Bruno his life.

While the moving, spinning Earth may have acted as a powerful stimulant to certain minds, the real challenge that lay before Europe's astronomers by 1570 was to find solid physical and mathematical evidence for the Earth's supposed motion, if Copernicanism was to stand as an actual description of how the heavens really were, as opposed to being simply the hypothetical 'phenomenon-saving' or calculating device suggested by Osiander. The way in which this was to come about was twofold. Firstly, astronomers had to measure and remeasure the heavens to determine whether any traces of a terrestrial motion, with relation to other astronomical bodies, could be detected. Secondly, the currently orthodox physics of Aristotle had to be undermined and discredited wherever possible, on the tacit assumption that, as every good criminal lawyer knows, the best way of defending a doubtful case is by undermining the evidence for the prosecution. Copernicanism gradually won a sort of covert credibility as one Aristotelian explanation after another was called into question.

The first person to make regular observations as a way of testing the physical credibility of the Copernican theory was a somewhat flamboyant Danish aristocrat named Tyge Brahe (1546–1601), although following the classical custom of the age, he Latinised his Christian name to Tycho. His eminent social background made it difficult for him to become an astronomer, because within the rigid hierarchy of Danish society, astronomers and academics were expected to come from the relatively small Danish middle class. It was an aristocrat's job, on the other hand, to govern and to wage war, just as it was a peasant's job to till the soil. At the age of fourteen, Tycho had been fascinated by the sight of an eclipse, and wanted to know how they could be predicted with such accuracy.

He was sent abroad, however, to study law and other subjects appropriate to a youth of his rank, at Wittenburg, Leipzig, and other leading universities. Yet his intellectual precocity, his social peculiarity, and his independence of mind, had already firmly asserted themselves, and by the time he was a twenty-year-old student at Rostock University, where he should have been studying law, Tycho was secretly teaching himself mathematical astronomy, and won early fame when, in 1566, he successfully predicted the death of Sulaiman 'The Magnificent', Sultan of Turkey, from an eclipse of that year. His achievement was somewhat dimmed, however, when it later transpired, in that age of slow communications, that Sulaiman had already died before the eclipse took place. In spite of his family's wish to obstruct his interest in astronomy, by his mid-twenties Tycho was coming to be recognised as one of Denmark's leading renaissance intellectuals, and while the doors of Copenhagen University were closed to him, his friend Charles de Dançey – the French Ambassador to the Danish Royal Court – let him lecture on astronomy and philosophy in the Embassy, to which Tycho attracted crowds of listeners.

Tycho's fame no doubt increased after a sword fight in December 1566, during which he lost the end of his nose. Nothing abashed, Tycho had a golden prosthesis made, which was probably secured to the inside of his nasal bone by means of a spring catch; and he was sufficiently proud of this feature to have the flesh-tinted nose included in all of his subsequent portraits. This unconventionality was further evident when, instead of marrying within his own aristocratic caste, as Danish law required, he formed an enduring common-law marriage with Kirsten Jørgensdatter, who was probably the daughter of the Lutheran pastor of Knudstrup. Only after Tycho and Kirsten had left Denmark and had gone to live in Prague (where the laws were different) after 1598 were they permitted to have their union legally acknowledged.

Needless to say, Tycho was not only fully conversant with the works of Ptolemy and Regiomontanus and the new Greek learning; he was also aware of the theories of Nicholas Copernicus, and had been since his teens. It is clear, indeed, that Tycho held Copernicus in the highest possible regard as a scientist, in addition to which one senses that he had something of a fellow-feeling for him because, along with Regiomontanus and himself, Copernicus was part of that cultural renaissance of northern Europe which was making yet further advances beyond those first made in Italy. It is also from Tycho, moreover, that we know what we do about Copernicus's main instrument, for in 1584, the then Canons of Ermeland (Frombork) Cathedral presented the great man's set of Ptolemy's Rulers to Tycho, who clearly felt honoured by the gift. But true to form as the astronomer who was destined to pioneer scientific error analysis, he set about quantifying the internal errors of Copernicus's 'Rulers' as a way of trying to apply corrections to some of his original observations. Tycho described and illustrated the Rulers in his *Astronomiae Instauratae Mechanica* (*The Instruments with which Astronomy was Reformed*) of 1598.

But what focused Tycho's energies as an original research scientist, and took him from being an eminent Danish intellectual to a scientist of international repute, was his work on the *nova stella* ('new star') which suddenly appeared in the constellation of Cassiopeia on the evening of 11 November 1572.

It had been believed since antiquity that the number of stars in the sky was fixed and unchanging, and while earlier Chinese and Arabic astronomers had recorded the presence of 'new' stars, and stars of variable brightness, none of this knowledge was available to sixteenth-century Europeans. It was all the more surprising, therefore, when the nova (which is now known to have been a supernova) flared up out of nowhere, and in a few hours became so bright as to be capable of casting shadows. Yet why was this star to play such a crucial role in the astronomical revolution?

In November 1572, no-one had any awareness of how damaging to the ancient philosophical background of astronomy this new star would turn out to be, for at first everyone assumed that it must be a meteorological body, or something in the upper regions of the air

amidst the Aristotelian element of fire, which was the supposed natural place for sudden changes in the sky. But since Tycho was already
coming to the realisation that the proof or disproof of the Copernican
theory must lie not in philosophical deduction but in the gathering of
fresh physical evidence, he resolved to make accurate measurements of
the position of the new star as seen in relation to the well-known stars
in Cassiopeia. To this end, he had built a large sextant, or instrument
capable of measuring angles up to 60 degrees, with which to observe
the new star at its rising and setting during the long Scandinavian
nights of the winter of 1572–73. Tycho worked on the correct
geometrical assumption that if the star really was in the Earth's atmosphere, it should show a regular shift – or parallax – backwards and
forwards between morning and night, in exactly the same way that an
outstretched thumb held out at arm's length will appear to move
against distant objects if looked at alternately with the left and the
right eye. But after several months of painstaking observation, Tycho
realised that the star showed absolutely no movement whatsoever. It
could not, therefore, be in the atmosphere, where Aristotelian theory
dictated it should be, but in *space*.

Then, in 1577 a bright comet streaked across the skies of the
northern hemisphere. Once again, since antiquity, comets had been
considered as atmospheric and not astronomical bodies: the products,
in fact, of 'effluvias' or foul stinks that had risen up from swamps or
from the unburied dead lying on battlefields, and had caught fire in the
air. Yet Tycho reasoned that if this was so, the comet of 1577, and other
comets, being in the 'sublunary' region between the Earth and the
Moon, should have easily measurable parallaxes from which their
distances above the Earth could be calculated. After all, the Moon itself
already had a well-determined parallax, and a fairly good value for the
lunar distance had been known since the days of Hipparchus in 150 BC.

Yet just like the new star of 1572, the comet of 1577 revealed no
measurable parallax whatsoever. Tycho was led to conclude from

these new and unprecedentedly accurate measurements that both the star and the comet were *not* in the atmosphere at all, but in space. Indeed, the case was strongly reminiscent of that regarding the geographical discoveries of a few decades earlier: a modern person, using a *physical* method of investigation, had shown that a centuries-old *philosophical* assumption was wrong. This suggested that the best advances in science and astronomy were made not by constructing elaborate schemes of logical cause and effect based upon a philosophical assumption, but by examining nature at first hand. And increasingly, from this time onwards, scientists were coming to think not just of pure Greek texts as the fountainhead from which all this knowledge sprang, but also of *Nature* herself.

By 1576, King Frederick II of Denmark realised that his Danish national asset needed more formal encouragement, and presented Tycho with the island of Hven, out in the sound just over the eastern horizon from Copenhagen. As a feudal lord in his own right, and with increasingly ample state resources at his disposal, Tycho now designed and built an observatory–residence named Uraniborg, or Castle of the Heavens; and to Uraniborg came scientific instrument makers to build the most accurate astronomical instruments that the world had ever known – all to Tycho's original designs. There also came young postgraduate students from across Europe. Most of these young men, it is true, came from the Germanic cultural orbit, but there were also Scotsmen and one young Englishman, John Hamon, or Hammond, of Kent and Cambridge, who may have been the same Dr John Hammond who later became physician to Sir Francis Bacon.

With its evolving instrument technology, professional staff, technicians, visiting students, and printing press, Uraniborg became Europe's first 'modern' scientific research institute. And like a modern scientific institution, it addressed itself to specific physical questions and studied them with carefully designed instruments, within the context of an overall research strategy. One of the things which Tycho wanted to do at Uraniborg was to observe the heavens afresh, to construct new tables of planetary motion rather than follow the medieval practice of simply overhauling Ptolemy, and to decide from a standpoint of factual evidence which of the two 'systems of the heavens' – the Ptolemaic and the Copernican – was physically true. To obtain the primary data from which this decision could hopefully be made, Tycho and his staff would use Uraniborg's new instruments to measure the angles between the stars and to minutely record the motions of the planets among them. In particular, Tycho hoped that by detecting certain kinds of regular changes in these patterns, he could find evidence for the Earth's motion around the Sun.

But alas, he found none. *Not*, of course, because these expected and searched-for changes amongst the stars and planets do not exist, but because they are in reality so very small that even Tycho's superlative instruments were too imprecise to reveal them. It would, indeed, take another 250 years of increasing sophistication in scientific instrument making before the tiny changes which Tycho hoped to detect would indeed be confirmed in the late 1830s.

On the other hand, Tycho realised that there was too much good sense in Copernicus for him to be wrong. In 1583, therefore, he proposed an ingenious system that combined the best explanations of both Ptolemy and Copernicus. Tycho proposed that the Sun was indeed the centre of rotation of all of the planets except the Earth. But then the Sun, carrying all of the planets with it, rotated around a stationary Earth in 365¼ days. On the one hand, Tycho's system made it possible to incorporate all of the elegant Copernican explanations for the retrograde loops, along with other phenomena, while on the other hand retaining Aristotle's physics of the fixed Earth. Tycho's theory was to have enormous influence in European astronomical thought over the course of the ensuing sixty years or so, and provided the perfect solution for those who were attracted to Copernicus's mathematical explanations yet were worried about his lack of direct observational evidence. In particular, as we shall see, it became especially important to early seventeenth-century Roman Catholic scientists.

In spite of his formidable stature as a scientist, however, Tycho never lost the imperious habits of a prickly aristocrat, and having angered and provoked jealousies among his fellow Danish nobles, he found that the new king, Christian IV, was beginning to curb his research revenues. Tycho responded by quitting his island (much to the relief of his tenants and peasants, no doubt, who found him a demanding overlord) and then Denmark in 1597. He was thereafter invited by the Holy Roman Emperor, Rudolph II, to reside with the Imperial Court in Prague, and he presented Tycho with Benatky Castle, to accommodate his household and to become his new observatory. But sadly, Tycho died suddenly, following an Imperial banquet in Prague. It seems that at the banquet, Tycho needed to make water, but protocol forbade him to leave the table. This water retention seems to have caused major organic damage, and on 24 October 1601 he died from its complications. Tycho, alas, died as unconventionally as he had lived.

By the time of his death, however, Tycho's name was famous across Europe, and not only were many people strongly attracted to his ingenious cosmology, but his thousands of unprecedentedly good observations of the positions of the stars and planets constituted a primary resource from which succeeding generations would conduct researches; and those researches, moreover, would cause astronomy to finally abandon some of its most venerated classical assumptions.

When Tycho died unexpectedly in 1601, he had several research assistants conducting mathematical analyses of his observations, with the purpose of using them to demonstrate the truth of his Earth-centred Tychonic system. One of these men was a clever 30-year-old German named Johannes Kepler. But Kepler's origins lay at the opposite end of the social scale from those of Tycho. His father had been a mercenary soldier, and his mother, on a later occasion, escaped a charge of witchcraft only after her distinguished son intervened. But Johannes – who always suffered from delicate health and poor eyesight – soon showed great intellectual promise, and he won a scholarship to the University of Tübingen, intending to train as a Lutheran pastor. But though the young Johannes never did become a Minister, he was, nonetheless, a devout Christian who, like many before him, saw his faith and his science as being inextricably linked, and the astronomical laws which he discovered as revelations of God's glory.

On the other hand, Kepler shared his age's fascination with the classical Greek roots of astronomy, and in particular with the Pythagoreans and their theory of number, geometry and harmony being at the heart of all things. It was almost certainly Kepler's Pythagoreanism, moreover, which from quite early on had convinced him of the truth of Copernicanism, and had produced in him a natural sympathy with the idea that the Sun and not the Earth stood at the heart of Creation. In 1596, when teaching at the University of Gratz, he suddenly realised that not only were the respective orbital periods of Mercury, Venus, Earth, Mars, Jupiter and Saturn in a harmonic sequence, but that the spaces between their still supposedly purely spherical orbits could be *exactly* filled by the five regular polyhedral solids (the cube, tetrahedron, dodecahedron, icosahedron and octohedron of classical antiquity) nestling inside each other. While his regular solids theory was later shown to be incorrect, it held within it a profound intuitive understanding of the relationship between speed, volume and spatial areas that a dozen or so years hence would lead him to radically change the postulates of Greek astronomy. But in the short term, it brought him to the notice of, and employment by, Tycho Brahe.

At the time of Tycho's death, Kepler was at work analysing the Uraniborg observations of the planet Mars, trying to find a geometrical mechanism whereby Tycho's observations could be made to work within the parameters of perfect circles and perfect speeds. Kepler obtained a much better agreement of results than he could have done using Ptolemy's system, but still found an inexplicable discrepancy of 8 minutes of arc between theory and observation. Now, for anyone's observations other than those of Tycho himself, a discrepancy of 8 minutes of arc would have been dismissed as the product of instrumental error. Copernicus's and Walther's observations, for instance, had generally been in error by 10 minutes of arc or more, due to the relatively primitive sights and scale divisions on their instruments. But Kepler knew that Tycho's observations were accurate to a *single* minute of arc, and sometimes less, so that a discrepancy of 8 minutes of arc represented something in nature that had to be explained.

Kepler, like Regiomontanus, was one of those individuals who delighted in calculation, and his work for Tycho necessitated his analysing hundreds of individual observations of the position of Mars and searching for numerical connections between them. Even after Tycho's death, and after his succeeding Tycho to the post of Imperial Mathematician to Rudolph II, Kepler still continued with the martian analyses, though now from a Copernican perspective. And then, in 1609, Kepler came to one of those conclusions that changed for ever the ground rules of astronomy; and between 1609 and 1619 he worked out its consequences to produce his famous three Laws of Planetary Motion. Kepler now abandoned the Greek prerequisites of perfect circles and uniform speeds. Instead, he realised that Mars moved around the Sun not in a circular but in an *elliptical* orbit, with the Sun closer to one end of the ellipse than the other. When Mars was closer to the Sun, it moved faster than when it was further away, although its velocity changed in an exact mathematical ratio to its distance.

From an analysis of the finest observations of a planet's changing position in the sky that had ever been made, combined with his extraordinary perception of how distances, shapes and volumes of space interrelated with each other, Johannes Kepler had shown that solutions to physical problems in science were not found by holding on to prior notions of philosophy and aesthetics, but by addressing nature direct. And while Kepler never lost his profound respect for classical Greek intellectual culture, he nonetheless realised that God had built a myriad of wonders into His Creation, and that scientists needed to be open-minded and willing to consider options of which the Greeks had been unaware, when they investigated it.

But the astronomer whose name came to be most closely associated with the astronomical revolution was Galileo Galilei (1564–1642), who has always been known to history by his Christian name. Galileo was the son of a Pisan lawyer, poet and musician, and after finding the dissection of human bodies distasteful, he abandoned a projected career in medicine, and turned instead to mathematics. By 1609, however, at the age of 45, now Professor of Mathematics at the University of Padua, near Venice, Galileo was feeling professionally unappreciated, hard up, and regretting the fact that fame had passed him by. Then, in May 1609, he heard from a correspondent in Paris of a device invented by a Dutchman which had the power of making distant objects appear close at hand. From his knowledge of optics, Galileo was soon able to construct a working 'perspective' or telescope, and promptly won some local fame, and a pay rise, when in August 1609 he showed a group of Venetian Senators how useful the device could be in wartime when looking at distant ships.

During the last few weeks of 1609, Galileo used the device to look at the Moon. Its surface, instead of appearing like smooth, albeit tarnished, silver, as the ancients had generally believed it to be, was found by Galileo to display distinct topographical features. These included mountains that cast such distinct shadows that when the Sun shone from the appropriate angle he was even capable of calculating their heights above the surrounding terrain. The dark regions of the Moon resembled earthly seas with distinct coastlines and islands, and today we still refer to these areas (now known to be lava flows) with the name which Galileo first gave them: *maria* (Latin – 'seas'). Galileo's lunar discoveries, moreover, seriously challenged orthodox classical Greek views about the nature of our satellite, for they suggested that the Moon, instead of being a perfect sphere, was in fact a rough and broken-surfaced *world*, just like the Earth.

In January 1600, Galileo discovered yet more astronomical dynamite when he saw that Jupiter was not just the brilliant star that appears to the naked eye, but that, when viewed through the telescope, it appears as a ball that is distinctly flattened at the poles. Yet most remarkable of all were the four little stars which he found revolved around Jupiter with the exact regularity of clockwork, and which he flatteringly called the Medicean Stars, after the Grand Ducal house of Tuscany from whom Galileo was hoping to win patronage. What made these Medicean Stars revolutionary, however, was their unmistakable rotation around Jupiter, for in doing so they were destroying the classical physical dictum that the Earth was the only centre of rotation in the Universe.

Galileo's first crop of telescopic observations – including those of the Moon and Jupiter, along with his discovery of dense star clouds in the Milky Way – were published in Venice under the title *Siderius Nuncius* (*Starry Messenger*) in March 1610. This little book was destined to become one of the most controversial publications in the whole history of astronomy, for Galileo used his discoveries to argue a radical intellectual agenda. In particular, he openly advocated the acceptance of the Copernican heliocentric theory, stressed the importance of new knowledge in science, and, most of all, he strongly criticised those conservative philosophers who dogmatically insisted upon the primacy of Aristotle's explanations for everything in spite of modern evidence to the contrary.

Unlike Copernicus, Galileo was an instinctive controversialist, for whom the techniques of argument and ridicule constituted a natural part of his armoury for advancing any cause. Indeed, long before the telescopic discoveries shot him to overnight fame across Europe, Galileo's controversial credentials had long been established, and one cannot overestimate the importance which his capacity for personal abrasiveness played in his subsequent relations with the Church.

In 1610, Galileo was at last enjoying the fame and renown he had so long felt he deserved. The Medici had responded most favourably, and Galileo left Padua and Venice – where he had never really been happy teaching mathematics – to return not only to his native Tuscany, but to the Grand Ducal Court. For Galileo not only now basked in the personal favour of Duke Cosimo II, but was also elevated to the prestigious Florentine Academy of Linxes, or 'sharp-eyed philosophers'. At the Grand Ducal Court, Galileo wore gold chains, dined with Cosimo, Cardinals and Ambassadors, and corresponded with the great men and women of Renaissance Europe. For the next couple of years, everything went well.

It was only really after Galileo's contested discovery of sunspots in 1612 that clouds began to glower on the horizon. Quite apart from the primacy of discovery, a much more obviously controversial issue arose regarding the very physical nature of the spots themselves. By 1612, Galileo was always on the look-out for ways in which he might advance the Copernican theory, and he realised that sunspots – just like the Moon's mountains and Jupiter's moons – provided him with yet more pro-Copernican ammunition. None of the newly discovered telescopic phenomena supplied a particle of physical or mathematical proof that the Earth rotated around the Sun, but they all undermined those conservative physicists who insisted that Aristotle was always right. It was Aristotle, after all, who had spoken of the Moon as smooth, the Earth as the only natural centre of rotation, and the Sun as a pure golden ball.

More conservative scientists – especially figures like the Jesuit astronomer Christopher Scheiner – interpreted the sunspots by describing them as being not on the Sun's actual surface, but rather as little planetoids passing around the Sun, and hence appearing in silhouette against the perfect golden light. In contrast, Galileo argued that the spots must be attached to the Sun, because they changed shape with location, appearing relatively round when in the middle of the disk, and diminishing to thin slivers at the edge. He then added a final *coup de grace* against the Aristotelians by saying that he had determined from the positions of the spots that the Sun rotated on its axis in 28 days, whereas the ancients had believed that not only was the Sun without any surface blemish, but that it stood fixed and unchanging, being devoid of any kind of axial movement.

It says a great deal for Galileo's perspicacity as a scientist that in every one of his radical new interpretations of astronomical phenomena, he was subsequently proven to be correct. Yet where his perspicacity failed him was in his instinct for controversy, point-scoring, and the delight which he evidently took in undermining his opponents. This trait did not greatly matter in the often acidic world of the Renaissance universities, where name-calling was frequently part of the nature of academic exchange, but it was different when such criticism was even implied against cautious and conservative Churchmen who were all too aware of how the Protestant Reformation had split once-unified Catholic Europe asunder, and undermined much of the Catholic Church's traditional prestige. And while those Churchmen, when wearing their lay hats, might have been sophisticated sons of the Renaissance, they were, when donning their bishop's mitres and cardinal's scarlet, too acutely aware of dangerous splits that had arisen within Christendom, and were adamant in preventing the further spread of what was seen as heresy.

One of the hallmarks of the new Protestant form of Christianity resided in the relationship between priests and lay people. Indeed,

Luther and the Protestant reformers objected to the spiritual preten-
sions of the Catholic clergy – who drew their authority from the
historical administration of the sacraments, along with their 'pipeline'
link back to Christ via St Peter – and argued that the Bible alone was
the ultimate source of authority, to which all men and women should
have access in their native language. The Protestant clergy, indeed,
defined themselves as no more than 'ministers' who might guide and
teach the people on the strength of their academic training, but above
whom they claimed no spiritual superiority. In consequence, therefore, and virtually from the very beginning, Protestant lay people had
taken an active role in their churches; and in some quarters, such as
England and Germany, by 1600 an exotic miscellany of tinkers,
tailors, serving-men and visionary women believed themselves theologians because they could stumble through a reading of the Gospels.

While Galileo was in no way influenced by these north-European
Protestant tendencies – living and dying a devout and obedient son of
the Roman Catholic Church – there were, amongst his utterances on
the new science, certain points to which an enemy could give a
dangerous twist. For one thing, certain Biblical passages spoke plainly
of the Sun moving in the sky, as when in the Old Testament Book of
Joshua (10:12) God had held back both the Sun and the Moon from
setting, to supply sufficient daylight for Joshua to rout the Amorites.
And did not *Psalm* 104 quite plainly state that God 'laid the foundations of the Earth, that it never should move at any time'? It could
therefore be construed as spiritually improper for a lay astronomer
such as Galileo to advocate the Copernican theory, as Father
Tommasio Caccini forcibly pointed out in the Roman church of Santa
Maria Novella on Advent Sunday 1614. While the Church had well-
established procedures for reconciling conflicts with Scripture (such as
the Apocryphal statement that only one seventh of the Earth's surface
was covered by water, the rest being land), it was the job of the
theologians and not of lay scientists to do it.

In 1615 Galileo wrote what was later perceived to have constituted an overstep of the mark, when in his published *Letter to the Grand Duchess Christina* he appeared to be teaching theology to the theologians. In this *Letter*, addressed to his patron's mother, Galileo advanced several arguments which have since become pivotal in the understanding of the relationship between Christianity and science. For instance, he argued that God had given *two* books to mankind: the Book of the Written Word (the Bible), which we read, and the Book of Nature, which we interpret by our senses (or by science). And as both Books are written by the same divine Author, and are by definition true, they cannot contradict each other, though as our knowledge of the Holy Spirit deepens, so what God had taught by a divine utterance to a more simple age, we might now discover in its full glory from a careful study of the Book of Nature. In spite of Galileo's now formidable intellectual standing and undisputed Catholicity, one can see here brickbats which priests who did not like Galileo's manner, or who may have been crossed by him, could use against him. Galileo was neither a priest nor a trained theologian, yet just like the radicals in Protestant Germany or England, he *seemed* to be trying to teach the theologians their business.

More urgent, however, was the problem about the truth status of the Copernican theory. Was the theory true as a physical fact (insofar as the Earth really did rotate around the Sun), or was it true simply as an ingenious hypothesis that facilitated planetary calculation? Most contemporary clerical scientists adopted the latter view, or else were supporters of the geocentric Tychonic system, which also facilitated easier calculation while keeping the fixed Earth. Galileo, on the other hand, was now openly arguing in favour of the full physical truth status of Copernicanism. The problem with Galileo's stance, however, was the admitted lack of observed physical evidence for the movement of the Earth. It was all very well for him to argue that the Copernican theory enabled better planetary calculation, and that sunspots and

Jupiter's moons undermined parts of the physics of Aristotle as a way of discrediting his opponents; but at the end of the day he could not produce a particle of evidence for the Earth's *actual* motion around the Sun. Indeed, the *experimentum crucis* for the motion of the Earth which Galileo himself was to advocate some years later – namely, the possible six-monthly parallactic displacement, or shift, of certain stars – remained unworkable until the time of Friedrich Wilhelm Bessel in 1838. Galileo had no idea how big the solar system really was, or how far away the stars were from the Sun; and consequently he did not know that stellar parallaxes are very tiny, and far beyond the detection capability of the instruments of his day. (The very largest stellar parallax known to astronomy – first measured in the skies of the southern hemisphere by the Scotsman Thomas Henderson in the 1830s – results in an angle which is the same size as that subtended by a 10p coin when viewed at a distance of 21 miles!)

As a result of his persistent defence of Copernicanism in the absence of firm evidence, and his partiality for ridiculing his more cautious opponents, Galileo had his first brush with the Roman Inquisition in February 1616. Following what seems to have been a very civilised discussion with the internationally eminent scholar Cardinal Robert Bellarmini, Galileo was forbidden to hold or teach the Copernican theory as a physical fact, although he seems to have been allowed to consider it as an *hypothesis*. On 26 May 1616 – as a way of silencing those among his rivals who were passing on the story that Galileo's Copernican views had been officially silenced by the Church, and that he had been formally punished for holding them – he obtained a certificate from Bellarmini stating the exact circumstances of the hearings.

Then, in 1623, Galileo's friend and ostensible well-wisher, Cardinal Matteo Barbarini, was elected Pope Urban VIII, and Galileo felt that his personal relationship with the new Pope might be used to lift the 1616 prohibition on advocating Copernicanism as a fact. But in spite of enjoying an easy access to his Holiness, and having many discussions with him, Urban would not lift the ban.

In 1625, after undertaking and publishing theologically innocuous though major researches into terrestrial physics, Galileo began the composition of his *Dialogue on the Two Great Systems of the World.* This work intended – in a supposedly open-handed fashion – to assess the merits of the Ptolemaic and Copernican theories, although it is noticeable that in his stress on only *two* world (or Universe) systems, he was omitting that of Tycho Brahe. This was clearly for polemical effect, for in 1630 the Tychonic system seriously undercut the explanatory power of Copernicanism, and Galileo, wanting a straight fight between what he perceived as a weak and a strong case, preferred not to muddy the water with cogent compromises.

By 1632, Galileo's *Dialogue* was ready for press. His friends and patrons in Florence liked it, and he set about the necessary process of submitting it to the Roman Censor to obtain his *imprimatur* to publish. Yet in his choice of a Censor, Galileo behaved rather heavy-handedly, approaching the genial and rather compliant Father Niccolo Riccardi – a high-ranking Vatican official who was not an expert on astronomy – to authorise the publication. The honest Riccardi was reluctant, but submitted to Galileo's arm-twisting; although he lived to regret it, for he subsequently got into serious trouble with the Pope. But in his bullying of Riccardi, whom Galileo clearly perceived as something of a soft touch, one sees evidence of that ruthlessness and determination to get his own way at all costs, which no doubt played a part in winning Galileo so many enemies. It also shows how extremely well connected he was with very senior officials in the Vatican hierarchy, and to think of Galileo as some kind of outsider struggling vainly against authority is quite simply wrong.

When the *Dialogue* was published in 1633 it was at first well received, especially in Florence and in Venice. Outwardly it seemed to be an open-handed discussion, set in the form of a debate between three men, a Copernican, a Ptolemaicist and an assessor. Yet one did not have to dig too deeply to find its real bias, and Galileo's enemies soon began to pounce upon them and draw them to the attention of the Pope. Why had Galileo chosen the name Simplicio (actually the name of a late Greek philosopher) for his defender of Ptolemy? And why did he – in the full flow of the imaginary debate between his protagonists Sagredo, Salviati and Simplicio – speak of those who opposed Copernicanism as 'idiots', 'mental pygmies', and even 'hardly human', when no such abuse was levelled against the Copernicans? But Pontifical fury truly broke loose when Galileo's enemies pointed out to his Holiness that some of the anti-Copernican arguments which Galileo had put into the mouth of Simplicio, indeed, included arguments that Pope Urban himself had used. It is difficult to imagine that Galileo, in spite of his love of polemical bravado, could ever have been so foolish as to deliberately poke fun at the Pope, who was both Christ's Vicar on Earth and Galileo's former friend. But as far as Galileo's Vatican enemies were concerned, it was all ammunition to use against the Florentine astronomer.

While the Pope was held to be Christ's Vicar, or representative on Earth, and as a mark of humility rode upon a white donkey or mule rather than a horse, most of the Renaissance Popes were men of power and pride – and one of the proudest of them was Urban VIII. He took little persuading, therefore, to authorise a trial of Galileo for having supposedly taught Copernicanism as a physical fact, in the very teeth of the injunction of 1616. But if Galileo was sometimes a difficult man to deal with, no-one could deny his sincere devotion either to Catholic Christianity, or to his search for truth, as opposed to hypothesis, in science, for he saw both as aspects of the same Divine Creator.

Galileo's trial before the Holy Inquisition in 1633 – polite and gentlemanly in its conduct as it seems to have been – was undoubtedly one of the biggest mistakes that the Catholic Church has ever made, for in officially silencing the elderly and ailing scientific celebrity, and sentencing him to house arrest at his villa at Arcetri, near Florence, it had sent out the very worst of all possible signals to the intellectuals of Europe. Where now were the Church's ancient traditions of scholarly debate? Where was Catholic Christendom's concern with *truth*, as opposed to dogmatic authority? And where was the humility of the man who was presumed to speak for Christ in the modern world and still kept up the pretence of riding on a simple donkey? Protestant scientists in England, Germany and Holland hero-ised Galileo as the victim of the Anti-Christ, and the French Catholic philosopher of science, René Descartes, left Paris to take up residence in Amsterdam.

The Galileo affair amply illustrated the dangers that could arise when a prominent and devout layman, whose researches were preg-nant with theological implications, presumed to discuss them in a world that was dominated by ecclesiastical princes. While it cannot be denied that, within their own spheres of understanding, both the Church and Galileo earnestly sought the truth, the post-Reformation Catholic Church felt itself to be in an embattled position, and was extremely wary of being taught its business by pushy laymen who *seemed* to be taking liberties of a kind more reminiscent of Protestant Wittenburg or London than of Italy.

It would, however, be incorrect to think of Galileo's condemna-tion as curtailing science in Italy and other Catholic countries. It is true that Copernicus's *De Revolutionibus* was put on the Index of Banned Books some ninety years after it had been written, and would stay there until 1835; but the Church quickly realised that in react-ing as it had to Galileo's *Dialogue*, it had shot itself in the foot. In spite of the Copernican prohibition, no other scientists were punished after Galileo.

Indeed, Italian science blossomed in its own way. Galileo's pupil, Evangelista Torricelli, continued his master's researches into the physics of moving objects, and in 1644 found experimental proof for the existence of the vacuum, which flatly contradicted Aristotle. The Florentine Accademia della Cimento pioneered experimental physics and telescopic astronomy in the decade after 1657, while in 1663 Marcello Malphigi used his microscope to find conclusive physiological proof for William Harvey's 1628 theory of the circulation of the blood in living creatures, in obvious contradiction to the classical medical writers. And none of these scientists were punished.

Curiously enough, however, astronomy also continued to flourish. The Jesuit Christopher Scheiner, while still believing that sunspots were not actually on the surface of the Sun, undertook a long series of observations of the solar surface, and recorded many hundreds of spot positions and changes between 1611 and 1626, culminating in the publication of his massive 1626 treatise on solar astronomy, which bore the cryptic title *Rosa Ursina* (*The Rosary of the Bears*). Likewise, Giovanni Baptista Riccioli and Francesco Maria Grimaldi, of the Jesuit College in Bologna, undertook important researches in the physics of swinging pendulums, using them to time a variety of astronomical phenomena; and also in Bologna, the wealthy aristocratic amateur astronomer, Count Cornelio Malvasia, maintained a large private observatory which undertook original lines of research and helped to train up young astronomers such as the later illustrious Giovanni Domenico Cassini.

In this post-Galilean world of Catholic science, moreover, it is impossible to overestimate the significance of that one Priestly Society or Brotherhood which, in spite of often opposing Galileo, nonetheless became the tireless promoter of the new science: the Jesuits. Nowhere were these energies more sustained than in their mission to China after 1620, during which the Jesuits attempted to use modern science, and especially the new calendrical astronomy, as part of a package of advantages that went along with conversion to Christianity. Fathers Johann Adam Schall von Bell and Ferdinand

Verbiest aimed at Christian conversion by starting at the top, attempting to win over the Emperor and royal household. During the 1660s and 1670s they almost succeeded with the young K'ang Hsi, who was an Emperor of a naturally intellectual disposition and was clearly fascinated by the instruments and machines that his Chinese-speaking Jesuit mentors had brought from Europe: clocks, telescopes, quadrants, modern firearms, cranes, pumps, volumes of mathematical tables, Greek and modern astronomical theories, and a harpsichord. Both Schall von Bell and Verbiest were successively promoted to the Directorship of the Imperial Astronomy Bureau, and Verbiest re-equipped the Peking Observatory with instruments based on Western prototypes. Their combination of sophistication, cleverness, flexibility and enterprise came within a hairsbreadth of converting K'ang Hsi, and one wonders how radically different subsequent world history would have been if by 1700 the Emperor of China, his court, and his nation, had become Christian.

But all of this Roman Catholic science was going on *after* Galileo's condemnation. Quite apart from the Church's reluctance to repeat the Galileo fiasco, Catholic astronomers quickly realised that if they simply avoided teaching Copernicanism as a physical *fact*, nobody in the Vatican would raise an eyebrow. If one used the right form of words – be it in the lecture halls of Bologna or the Imperial Court of Peking – modern ideas could be explored, and when K'ang Hsi was shown the moons of Jupiter and the Galileo-discovered phases of Venus through Verbiest's telescope, they were generally discussed from a Tychonian-cosmology point of view. And so it continued until, in the nineteenth century, the Roman Catholic Church began moves towards publicly admitting that it had acted both unwisely and wrongly in 1633 (as we shall see in Chapter 12).

Europe's renaissance astronomical revolution was undoubtedly one of the great watersheds in the history of human thought. Beginning from a position which assumed a conservative reverence for the Greeks, it had unexpectedly discovered whole areas of new knowledge that showed the Greeks to be simply human and not omniscient after all. One salient feature of this process of discovery, in which the denizens of the last age of the world had shown themselves to be every bit as clever as their ancient ancestors, was the methods whereby they had discovered their new knowledge. Had not the Greeks been wrong in their insistence that only knowledge gained by logical deduction, geometry, or pure reason was truly valid? Had not the men of the modern age shown that empirical knowledge, gained by driving ships across hitherto unknown oceans or discovering stars invisible to the naked eye, was equally valid when conducted within a critical context of proofs and controls? Who, indeed, could deny that ships, Tycho Brahe's unprecedentedly accurate observatory instruments, and Galileo's telescope, extended and refined the human senses and opened up new oceans of truth before them? And who, moreover, could deny that if, as Galileo had suggested, God was constantly revealing new pages of His Book of Nature to those creatures whom He had made in His own image, He was not only our Maker, Redeemer and Friend, but also our continuing Schoolmaster?

ELEVEN

*From the falling Chancellor
to the flying bishop: astronomy
in the English Renaissance*

That cultural phenomenon which we think of as the European Renaissance flowered in different countries at different times, and often in different subjects. Italy's genius had lain in recovering the Greek and pure Latin (as opposed to medieval or 'dog' Latin) tongues and their respective literatures, along with architecture, painting, landscape design, navigational expertise, theatre, political thought, and the philosophy of science. Germany, by contrast, had found its primary cultural flowering in theology and music: in the shaping of those spiritual forces which had produced the Reformation and had gone on to give voice to a people by enabling them to offer up their praises in the great chorales and organ works of Hans Leo Hassler and Johann Sebastian Bach. England's Renaissance, when at its most original, however, was very much about physical science: about explorer–patriot sea captains, such as Sir Francis Drake and Sir Walter Ralegh, and about men whose approach to science focused less upon the thoughts of the ancient Greeks and more on the development of a disciplined system of experimental knowledge.

From the 1560s onwards, in fact, these emerging tendencies had already started to become apparent in English intellectual culture. During the 1550s, Tudor England's 'Renaissance Man', Dr John Dee, had already set minds ablaze in Cambridge, Louvain, Paris and Rheims with his lectures on mathematics and the Platonic power of numbers, while his 'Mathematical Preface' to Henry Billingsley's English translation of Euclid's *Elements of Geometry* (1570) had begun to point towards a mathematical understanding of nature. It was Dee, moreover, with his parallel interests in Angel Magic and the occult, who had formed the minds of a whole generation of brilliant young Englishmen during the reign of Queen Elizabeth I, while even Her Majesty enjoyed conversations with Dr Dee, and openly encouraged him in his researches. Thomas Digges was the first individual to write an English-language account of Copernicus's ideas and propose a model for an infinite Universe (as we saw in Chapter 10), and had been a pupil and protégé of Dee, following the death of his father Leonard Digges, who had also been a close friend of the Doctor. And in 1571 – long before the telescope made its formal entry into European consciousness in 1608–09 – Leonard Digges left a cryptic account of an optical instrument employing mirrors which could, allegedly, make distant objects appear close at hand.

Dr Dee would also influence the young Thomas Harriot who, sailing under Sir Walter Ralegh's patronage in 1585, left the first scientifically accurate account of the people, topography and natural resources of a part of the North American seaboard, which he published in 1588 under the title *A Briefe and True Report of the New Found Land of Virginia*. Twenty-three years later, Harriot, using a newly acquired 'Dutch spyglass', would become the first person to see sunspots and draw a map of the Moon in advance of that of Galileo. For while Harriot, like Galileo, was a convinced Copernican, he was a well-to-do bachelor who shunned publicity and hated controversy, so that his telescopic discoveries only came to light many years after his death when his private papers were being examined for the first time.

Also connected to the Elizabethan Court was Dr William Gilbert, one of the royal physicians. Queen Elizabeth was uncommonly healthy for a sixteenth-century person, and did her best to avoid doctors. Gilbert must therefore have had plenty of time on his hands, and he devoted more than twenty years to the study of magnetism. In addition to his discovery of the north and south poles of magnets, along with the directions of their lines of force, Gilbert experimented upon spherical magnets, or *terrellas* (little Earths), and even developed an argument in favour of the Copernican theory that derived from the Earth being a spinning magnet. It is also probable that Gilbert's book *De Magnete* (*On the Magnet*), published in 1600, subsequently influenced the cosmological ideas of Johannes Kepler, for at one stage Kepler suggested that the invisible force that kept the planets orbiting around the Sun might be a magnetic flux emanating from a spinning Sun.

It is surprising, indeed, how these and other Englishmen before 1600 had been clearly influenced by the writings of Copernicus, and would go on to influence others in their turn. And then, in 1597, Gresham College was founded in the Bishopsgate Street, City of London, mansion of the late Sir Thomas Gresham, the great Tudor financier, who had left his wealth to the City of London to establish a remarkably novel form of academic foundation. Gresham College, with its seven endowed resident bachelor Professors, was created to provide public lectures in the Seven Liberal Sciences to the citizens of London, and visitors, free of charge. But the College never had undergraduates or even registered students. Instead it was perhaps the world's first college of higher adult education, where anyone could drop in to attend lectures on astronomy, geometry, and so on. To emphasise the College's international as well as its metropolitan credentials, the Professors were required to give their lectures first in English, for Londoners, and then to repeat them in Latin, for the benefit of foreign visitors.

Edward Brerewood, Henry Briggs, Henry Gellibrand and many other early Gresham Professors were astronomers and mathematical scientists of the highest calibre who, in addition to their researches into geomagnetism, instrument designing, and astronomical observation, worked closely with explorers and navigators as a way of making their knowledge of practical value – especially to the Merchant Venturers of London – for cartography and safety at sea. And then, in 1619 Sir Henry Savile, Warden of Merton College, Oxford, established two new Professorships, one in astronomy and the other in geometry, to further propagate and encourage these disciplines at Oxford University, so that by 1640, the Gresham and Savilian Professors were active in teaching and conducting research into the new sciences.

These were some of the ways, therefore, in which Tudor England responded to the Renaissance – while at the same time producing a spectacular literary flowering which culminated in the sonnets and plays of William Shakespeare and the Authorised Version of the English Bible, to which King James I gave his imprimatur in 1611.

There was in this world, however, a figure whose impact upon the growth and direction of scientific culture is impossible to overestimate. This was Sir Francis Bacon (1560–1626), a lawyer, scholar and courtier whose brilliant yet ultimately ill-starred career could well have formed the stuff of a Shakespeare tragedy, had not Shakespeare predeceased Bacon by ten years. Bacon's career was ill-starred in the respect that his youthful and sustained brilliance took forty-five years to rise out of the doldrums of back-bench parliamentary politics and the grind of hired work in the law courts, to a sudden and rapid ascent through the ministerial ranks, to the gilded office of Lord Chancellor (which in those days was closer to the modern office of Prime

Minister) in 1618. Why Bacon's career languished during the long reign of Queen Elizabeth I – who when young had dandled the infant prodigy on her knees – is not clear, though some of his early criticisms of government taxation policy, combined with Sir Francis's marked preference for handsome young men probably played a part. But when King James I ascended the throne in 1603, Bacon's career took off, and one suspects that the new King's own partiality for handsome young men removed an old obstacle against Bacon's preferment. After he had risen to the supreme height of the Lord Chancellorship, Bacon's many enemies – for he was also high-handed, notoriously arrogant, and now inclined to favour the King instead of Parliament – engineered his fall on a minor yet undeniable charge of corruption. Although King James could not, for probity's sake, avoid sacrificing his Lord Chancellor, his Majesty did commute all the punishments, so that Bacon could retire to private life. According to his near-contemporary biographer, John Aubrey, Bacon met his end from the complications of a chill which he had caught when stuffing a chicken with snow, while performing experiments in refrigeration. King James raised Bacon to the peerage with the title Baron St Albans (Latinised *Verulam*, the Roman name for St Albans) and one wonders what might have gone into a play by William Shakespeare entitled *The Tragicall Historie of the Rise, Fall, and Untimely Death of my Lord Verulamium*.

One aspect of Bacon's youthful brilliance – even when he was still a very young undergraduate at Trinity College, Cambridge – had been his recognition of the aridity to which the teaching of Aristotle had descended by the 1570s. For as Bacon recognised, Aristotelian philosophy was essentially about *words* or logical categories rather than about *things*. Earth, fire, weight, perfection, colour, and so on, were labels that one stuck upon natural phenomena, but they were not explanations. How, for instance, did one obtain a better understanding of the dynamics of falling bodies by simply saying that things fell to mother Earth because they were innately heavy? On the other hand, one might ask how an ambitious barrister and would-be courtier became so involved with science that he is generally thought of as the father of experimental method.

Three things, one might argue, point towards the answer. The first was his fascination with truth and how one might unearth it. The second was his long exclusion from his much-desired access to political power, which exile gave him the freedom and mental space to ponder deep questions. And the third was the very nature of English legal training and practice in Bacon's day.

Bacon's love of truth shines through his writings, and it is clear that his ponderings and researches had probably begun in the early 1590s when he was a working barrister and Member of the House of Commons. For unlike such persons today, Tudor lawyers and MPs often had plenty of spare time in which to think great thoughts. The genesis of Bacon's philosophy dates from this period, and he came to envision a Reformation in learning in a similar way to which there had been a Reformation in religion. At the heart of this Reformation must lie a questioning of customarily held academic opinions, for, as Bacon was at pains to argue, the world of learning worshipped blind ideas, many of which stemmed from an excessive reverence for Aristotle. Instead of sticking 'wordy' labels on heat, cold or magnetism, one should analyse nature with an open mind, observe, collect examples of phenomena, and classify them. Bacon advocated a thoroughgoing empiricism in science, which worked on the assumption that the human senses, when operating under stable and controlled conditions in conjunction with an open yet cautious mind, could discern all sorts of new phenomena. These phenomena, moreover, once collected, should be carefully arranged into contextual empirical schemata, and from them the truths of nature would emerge. As one experimental and observational truth built upon another, this mode of inquiry became *progressive*, and led to the *Advancement of Learning*, which was also the title of Bacon's treatise of 1605, which really began his 'Great Instauration', or revival of learning.

Yet what has this to do with law? Well, four centuries ago the bulk of England's 'Common Law' lay not in Parliamentary Statutes, but in accumulated previous courtroom decisions, or case law. A law student, to prepare for his 'calling to the Bar', had to memorise scores if not hundreds of legal decisions and precedents, and be able to recall and argue their salient points when being examined. This made English law intensely practical and pragmatic in character, hinging as it did on exact points of evidence rather than upon *a priori* theoretical principles from which decisions could be logically derived. Cross-examination and argument also lay at the heart of legal procedure, especially in criminal law, so that a law student, when he qualified, thought instinctively in terms of looking for and exploring an opponent's faults, in a legal system that was adversarial rather than administrative.

When one looks at public or state law in Bacon's time, moreover, one finds an obsession with treason, the sniffing out of cunning Roman Catholic plots, and with people not being what they seemed. Bacon himself had appeared for the Crown Prosecution in the case of the Earl of Essex's treasonable insurrection in 1601, while the Gunpowder Plot, in which Guido Fawkes and a group of Roman Catholic conspirators had tried to blow up the King in Parliament on 5 November 1605, has since entered into national legend. If one was cross-examining a suspected conspirator, and trying to force him to incriminate himself and his co-plotters, very brutal means of persuasion – such as torture in the Tower of London – could be employed. And from a network of forced confessions, the prosecuting lawyer would build up the big picture of the whole crime, arrest all those involved, and identify those foreign backers and paymasters who were trying to overthrow the Protestant English state.

Bacon knew this world intimately, and from what we know of the legal priorities of Bacon's age, we cannot help but be struck by how his ideas on science often parallel those of law. In addition to his scepticism about Aristotle and academic idols, the intensely fact-driven, anti-speculative and taxonomic character of contemporary legal evidence probably had a formative influence in his empirical, experimental and tabulated approach to science. Bacon also clearly thought of 'nature' as a reluctant witness out of whom secrets could be prized by the appropriate forensic techniques, and his characterisation of the experimental method in science as 'putting Nature to the torture' is distinctly redolent of the Tower of London and the rack.

Legal analogies and the torture chamber were not, however, the only formative influences on Bacon's scientific thought. There were also the great geographical discoveries of the previous century. Columbus, Magellan, Drake and the English navigators also figured prominently in Bacon's scientific imagination, and many references to their achievements occur in his writings. They were influential in Bacon's thought because they so clearly broke loose from the bonds of classical prescription, and showed those men of the present age who still blindly followed Aristotle that there were great wonders out there if one only had the courage to grasp them. One also suspects that the nakedly empirical character of geographical discovery, whereby hard fact overturned theory, appealed to Bacon's lawyer-like mind. It was surely not for nothing that the allegorical frontispiece to his *Novum Organum* (*New Method*, 1620) showed a modern ship under full sail passing between those two Pillars of Hercules which had marked the self-acknowledged limits of ancient geography, and on to the great ocean of truth that lay beyond them. In the posthumously published *New Atlantis* (1628), Bacon used the fictional device of a group of sailors discovering a hitherto unknown island, the whole advanced civilisation of which was devoted to scientific research and its application to the common good.

Yet Bacon did far more than simply think and write about science. He also used his wilderness years to practise it. And while astronomy was not one of his favourite lines of investigation – probably because the science was, by definition, so untactile – he did work in physics. The properties of heat, cold and magnetism especially interested him, and he left recorded investigations into these subjects. His research on heat and friction led him to conclude that heat – or what Aristotelians called 'fire' – was not an element, but the product of physical motion. Modern physicists would agree with Bacon.

One might also suggest that another trait of the lawyer's mind – reinforced by his student experiences in mastering the great digests of case law – was important to Bacon's science: his historical perspective. While Bacon saw science as being capable of transforming the future, it did so by identifying tendencies which had moulded its past. The modern world of 1600, Bacon argued, was made possible by three great inventions: gunpowder, printing, and the magnetic compass. Between them, they had transformed the limited technical capacities of the ancients and made possible new forms of warfare, intellectual propagation, and scientific navigation. All of these, moreover, were examples of the Greek *techne*, or the power of technology, which was more than simple handicraft, but a product of the disciplined control of the physical world by the application of the mind.

This *techne*, moreover, could be systematically expanded and applied to new areas of enquiry, which Bacon especially designated as the improvement of agriculture and navigation, and the 'Prolongation of Life', or scientific medicine. All of this, moreover, must be done from noble motives. True scientific research must be 'Luciferous', or driven by a wish to find Light (Latin *lux*) as its leading motive, and not 'Lucriferous', or concerned only with lucre or personal greed. Even so, by pursuing the divine light of pure knowledge and gaining wisdom, scientists could also add to the sum total of prosperity and material security which made the 'Commonwealth' of England strong and happy. For wisdom begat ingenuity, and in their turn they laid the necessary foundations for abundant manufactures, trade, and peace.

Watches, firearms, glass windows, coal, sugar, plentiful iron goods, and the abundant fruits of the Indies and the Americas, were the exotic articles of commerce in this new world – articles of which the ancients, indeed, had been very largely ignorant.

Bacon also saw this expansion in a profoundly religious context. As a committed Protestant, and the son of a possibly Puritan mother, he was all too well aware of God's providence to His Creation, and of mankind's duty to seek out and make proper use of God's bounty. This attitude, moreover, found its finest expression in another of those elegant and pithy phrases which characterise Bacon's thought and writings: we were behoven to pursue scientific research 'For the Glory of God and for the Relief of Man's Estate'. Never, since Bacon wrote this aphorism in the early seventeenth century, one might argue, has anyone devised a finer expression of the relationship that should exist between pure and applied science.

England's receptivity to the new science in the early seventeenth century was also important in that gradual corrosion of belief in the authority of astrology which had been present in Western thought since earliest antiquity. Over the centuries, several things had happened that had weakened astrology's credibility, and one of the most powerful of them had been the Christian faith. Astrology's original credibility had hinged upon the planets being 'gods in the sky', or supernatural agencies that had some kind of influence upon the minds and bodies of mortal humans. Yet the essential monotheism of Christianity – expressed in the Trinitarian unity of the Father, Son and Holy Ghost – banished by definition any polytheistic intelligences that might have any independent psychic jurisdiction within God's Creation. By its very claims to predict the future, and therefore somehow curtail that free will which God had given to His human creation, astrology was seen as essentially heretical.

Even so, astrology had been openly practised during the medieval and Renaissance centuries – even by Church men for, as they sometimes argued, the celestial geometry of the horoscope did not actually bind the will, whereas the astrological art could be useful in providing warnings on health and similar matters. It was in medical diagnostics and prognostics, indeed, that astrology was perhaps most widely used in Tudor England, working on the assumption that different zodiacal types – such as people born under Leo or Scorpio – were inclined to different illnesses. For if each star sign had a human 'type', and each of the seven planets corresponded to a specific set of human attributes (such as Venus inclining to lust, and Saturn to melancholy), then the relationship existing at any time between the planets and the signs could provide useful warnings.

Yet quite apart from any ecclesiastical objections to the morality of astrology, in the seventeenth century two major changes in scientific ideas came that knocked away the intellectual and physical foundations upon which astrology had traditionally stood. The first of these was the Copernican Revolution and its consequences. The ancient plausibility of astrology had stood four square upon the 'fact' that the Earth occupied a special place in creation: fixed and stable at the centre of the Universe, with all the planets and stars rotating around it.

However, after Galileo's discoveries in particular, a growing body of evidence suggested that the Earth was moving around the Sun. Galileo's telescope, moreover, had shown that instead of being mysterious or transcendent bodies, the Moon and Sun had blotched and (in the case of the Sun) *changing* surfaces. The planets were then not just bright lights, but showed themselves to be spherical worlds with peculiar characteristics of their own: Venus exhibited phases like those of the Moon, depending upon her position in relation to the Sun; Jupiter looked like a slightly squashed orange, and had four moons all of his own; and Saturn was a ball which sometimes appeared with a pair of peculiar round projections, one on each side, which he seemed to eat up and disgorge at regular intervals, in the same way that the

Greek God Kronos (Saturn) devoured his own sons! (In 1657, the superior telescopes of Christiaan Huygens, in Holland, would reveal these 'projections' to be a broad, flat ring that surrounded the planet and which could be seen to appear and disappear with its changing axial tilt.)

But the second body of new knowledge which fundamentally undercut the intellectual credibility of astrology in the seventeenth century was medical. The planets were traditionally said to act upon a person by changing the balance of their four humours – yellow bile, black bile, blood and phlegm. Mars in Scorpius, for instance, could produce violent sexual behaviour by further inflaming the body's already hot yellow bile. As the seventeenth century wore on, however, the ancient doctrine of the humours fell increasingly out of serious medical purview. Harvey's discovery of the circulation of the blood in 1628, Robert Boyle's and John Mayow's work on the chemico-physiological basis of respiration in the 1660s and 1670s, and an increasingly sophisticated knowledge of anatomical structures, all helped to incline scientific opinion towards a *mechanical* theory of the body and away from a *quasi-occult* theory. And as the classical humours faded away as serious physiological agents, so their responsiveness to celestial forces also faded. Astrology slipped out of informed opinion, therefore, not because of Church or any other form of persecution, but because the old geocentric cosmology and the medical doctrine of the humours on which it depended as a causal system ceased to be believed in by educated people.

Not only did astrology suffer a major body blow when educated opinion came to accept the weight of evidence in favour of the Earth's rotation around the Sun, but at the same time the very nature of the ancient constellations was also coming to be challenged. For whenever one looked into space with a yet more powerful telescope, one saw not only the familiar bright stars, but more and more dim stars. Could it be that instead of the stars all being the same distance from the Earth and forming real constellation shapes upon the inside of a black sphere, as the ancients had argued, they were in reality scattered infinitely in all directions throughout space? This is exactly what was suggested by the new evidence gathered by increasingly powerful telescopes.

Yet if the stars were in reality scattered throughout an infinitely vast Universe, and the zodiacal constellations had no more relationship with each other than did the bright and dim flickering lights of a great city seen at night, then what sense did it make to speak of Taurus, Virgo or Aquarius? Instead of being geometrical and perhaps spiritual entities in their own right, these constellations were nothing more than mere chance line-ups of separate stars. In spite of his own interest in astrology, Thomas Digges had begun to grasp these possibilities as early as 1576, and Galileo's English disciples after 1610 would hammer them home.

In addition to metropolitan and Oxbridge Copernicans such as Thomas Harriot, Henry Briggs and others, however, it was in Lancashire and Yorkshire that Galileo's and Kepler's discoveries found their most fertile and creative soil in England after 1630. In this respect, three young men in particular were of especial significance: William Crabtree, a Salford (near Manchester) cloth merchant, and his friends Jeremiah Horrocks, a young Cambridge graduate, from Liverpool, and William Gascoigne, a Roman Catholic gentleman from Middleton, near Leeds. Although Horrocks and Gascoigne had been to university, where they would no doubt have become acquainted with Ptolemy and Sacrobosco in the *Quadrivium*, their well-documented familiarity with the 'new' astronomy of Galileo, Tycho and Kepler would, like that of Crabtree, have arisen from private study.

The surviving correspondence of these three remarkable men between 1636 and 1643 heralds the beginning of fundamental astronomical *research* in Britain, insofar as they did not merely imitate the works of the continental masters, but went beyond them. They ridiculed astrology as groundless, and they evaluated the Copernican and Tychonic theories and came out on the side of Copernicus. They used Tycho's observations – published in Kepler's *Rudolphine Tables* (1627) and elsewhere – and then checked their tabular predictions against their own observations. As a consequence, they found errors, which they corrected by using Keplerian models based on elliptical orbits. Horrocks' work on the elliptical orbit of the Moon was of fundamental importance in the history of celestial mechanics, and won warm acknowledgement from Sir Isaac Newton in his *Principia Mathematica* (1687), supplying as it did some important concepts towards the development of the theory of universal gravitation.

But the work for which Horrocks and Crabtree came to be immortalised in the history of astronomy was their successful prediction and observation of the first witnessed transit of Venus across the disk of the Sun, on 24 November 1639. Even Kepler had failed to predict this event, whereas Horrocks was able to use the observation to make, amongst other things, key corrections to the then imperfectly understood orbit of Venus. These observations then enabled him to posit that Venus, like Mars, probably moved around the Sun in an elliptical orbit.

In addition to their observational and mathematical achievements, this small group of north-country friends made major contributions to the technology of astronomical instrumentation. In 1640 William Gascoigne found a solution to an optical problem which had baffled Galileo, Kepler and other continental giants: namely, how to place a cross-hair or other marker in the field of view of a telescope so that the instrument could be used to make precise measurements. Soon afterwards, Gascoigne went one stage further and invented the micrometer, in which the threads of a delicate screw enabled the user to make angular measurements of unprecedented accuracy.

Young as these men were in the late 1630s – between 20 and 28 – their careers were short-lived, for disease and civil war battles were to see them all dead by 1644. Even so, in the late 1710s, when the then elderly Revd John Flamsteed (the first Astronomer Royal) was writing a history of astronomy in the Preface to his own life's work, *Historia Coelestis Britannica* (*British Account of the Heavens*), he was unequivocal that these young men were the true founding fathers of English astronomical research, and the English heirs of the continental giants.

A sense of the speed at which British astronomy was moving by the 1640s can also be appreciated when considering another young Lancashire man, Jeremey Shakerley from Pendle, who was allowed to read the recently deceased Horrocks' and Crabtree's unpublished manuscripts, then preserved at the Burnley mansion of Richard Townley. Until around 1647, Shakerley had been keen on astrology, but Horrocks' papers changed his mind entirely. He now became an 'evangelical' Copernican, and quickly saw the inadequacies of astrology when viewed from the new perspective. Then, in either 1650 or 1651, he travelled to Surat in India, from where he sent back observations of comets, a telescopic observation of the 1651 transit of Mercury across the Sun's disk (in the wake of Horrocks' transit of Venus), and some early accounts of Brahmin astronomy. And it was in India that Shakerley's scepticism regarding astrology became complete, for, as he pointed out, the long-familiar meteorological astrology of Europe did not work in India. Why, for example, if the same configurations of planets shone down from the same sky, did India and England have such different storm and rainfall patterns, when both countries were on the same Earth? Could it be that Indian monsoons and wet English winters were caused by *natural* forces that had nothing to do with horoscopes?

One major consequence of the continental astronomical revolution that was very much in the minds of the English Copernicans was the calling into question of the crystalline spheres upon which the planets were allegedly carried. Since Tycho's discovery that the new star of 1572, and the comet of 1577 and others seen during the 1580s, were in astronomical space and not in the Earth's atmosphere, questions had been raised about the very existence of the spheres and the nature of space. Could Tycho's comets, moving in their strange orbits, somehow be crashing through the spheres? But the real *coup de grace* against the ancient spheres really came with Kepler's work on the elliptical orbit of Mars, for it was now stretching classical explanations beyond endurance to pretend that Mars and the other planets moved at variable speeds around a set of transparent crystalline *eggs*.

Yet if space was seen as empty – quite literally *space* – then what was the force that kept the newly discovered spherical worlds spinning so perfectly round the Sun? Kepler – no doubt influenced by Gilbert's *De Magnete* and its announcement of the existence of invisible magnetic fields of force – had suggested that the Sun could perhaps be an immense magnet rotating on its axis, and emitting a massive force-field into space. If this was the case, indeed, it would conveniently explain why planets further from the Sun moved more slowly in their orbits: Mercury – the planet nearest to the Sun – completed its orbit in 88 days, while Saturn – the furthest then known – took nearly 30 years. In the same way that an ordinary magnet's effect upon the needle of a compass became weaker with distance, so the solar force was no doubt weaker by the time it reached to Saturn.

The 20-year-old Jeremiah Horrocks (at the time when he was working most probably as a schoolmaster and Church Bible Clerk in the Lancashire village of Much Hoole, near Preston, in 1639) was also very much concerned with the question of how a spinning Sun could make the planets move in a Universe without crystalline spheres. In his subsequently published treatise on the transit of Venus in 1639, for instance, he discussed the possible force that might be responsible for making Venus and the other planets rotate around the Sun, considering that 'natural and magnetical causes' might be acting. At the same time, Crabtree was advocating 'Magnetical or Sympathetical Rayes', after Kepler.

Although Horrocks and his friends mention René Descartes (who outlived them all, dying in 1650) in their letters, it would be interesting to know in more detail how familiar they were with the work of this great contemporary French philosophical scientist. Descartes was also wrestling with the problem of how the planets could move in a Copernican, Keplerian Universe which was empty and devoid of spheres, although his solution was an integral part of his much broader philosophy of nature, which sought to explain how things act and move in the post-Platonic and post-Aristotelian world of early seventeenth-century Europe. By instinct, Descartes was a mathematical physicist and logician who believed that the golden key to the whole of scientific knowledge lay in finding a coherent explanation for why physical objects move and remain in motion. His answer was to see the whole vastness of God's Creation – from the motion of blood around the human body to the orbits of Jupiter's moons – as part of a grand piece of mechanism. Things moved because some antecedent motion pushed them, like the gear teeth of a watch. It was for this reason that Descartes was regarded as the founder of the Mechanical Philosophy.

This concept of seeing nature as an all-encompassing system of moving parts possessed great explanatory potential. Living creatures, for instance, moved because their hearts pumped nutritious blood, their bones and joints were mechanical levers, and their muscles were reflexive springs. And planets moved because the Universe was filled, Descartes hypothesised, with a boundless infinity of minute *corpuscles*, or particles that occupied every space, everywhere. These corpuscles, moreover, were in ceaseless motion – set going at the beginning of time by God – and were rotating in great and small whirling *vortices*.

The planets moved, therefore, because each was carried in its own respective vortex around the Sun. Comets described strange orbits because they passed between adjoining vortices, while rays of light were carried on their journeys from the stars to us by the most subtle or delicate of moving particles. Of course, Descartes could not provide a physical proof for the existence of his vortices; but they supplied a working hypothesis that was more successful at explaining the physical phenomena newly discovered in the early seventeenth century than were the four elements of the ancient philosophers. And by stressing the importance of the observation and measurement of moving objects, Descartes' Mechanical Philosophy played a major part in explaining nature in physical and mathematical terms, instead of in old animistic and quasi-occult terms.

Yet Descartes was all too well aware of the potential religious problems of his thoroughgoing mechanism; for if our bodies are but self-acting machines, how do our immortal souls fit in? Over the years, Descartes and his disciples proposed several solutions – such as the human soul being lodged in the cerebral pineal gland, where it could somehow interact with the physical brain and with our perceptions – though all of these were ultimately found to be unworkable.

What Descartes had done, however, was identify for the first time in the history of science that concept which we now call the 'mind–body problem'. This problem is still with us, and while modern-day neurologists possess an incomparably more sophisticated knowledge of the workings of the brain and central nervous system, we are, in effect, no nearer to an adequate or satisfying solution than was Descartes in 1650. For while modern knowledge of brain function can tell how we *perceive*, it cannot tell us why we *desire*, why we are *conscious*, or why we possess a sense of *selfhood*. Ultimately, as Descartes realised, this is where two different types of existence – the physical and the divinely transcendent – meet together.

But when we consider such issues as the mind–body problem, the Copernican theory, the infinite Universe, the causes of planetary motion, and numerous other issues that were under discussion by

1650, we obtain an idea of how far science had come since Columbus discovered America a mere century and a half earlier. Indeed, more original scientific knowledge had emerged during that period than during an equivalent stretch of time in the most golden years of Greece. The moderns had shown themselves to be every bit as clever as their classical masters – and perhaps nowhere more so than in astronomy and cosmology.

It was also during those heady years of the first half of the seventeenth century that what we think of as two very modern scientific concerns found their first coherent expression: life on other worlds, and space travel.

Although the possibility of an inhabited Universe had been raised as far back as the fifteenth century, the serious discussion of the idea was a natural offshoot of Copernicanism and Galileo's telescopic discoveries; for if the planets were now known to be *worlds* with continental and topographical features similar to those of our Earth, why should God not have given them inhabitants? And in the mechanism-bedazzled early seventeenth century, when new discoveries and inventions flew so thick and fast that Jeremy Shakerley was to observe that 'no day passeth without a triumph', a number of men were seriously wondering how long it would be before some Columbus of space finally placed his feet upon the Moon – and beyond!

Around 1630, an astronomer of considerable fame – Johannes Kepler – wrote the world's first science fiction story: *Somnium* (*The Dream* – published posthumously in 1634). In accordance with the best astronomical and physical knowledge of the day, Kepler told of a young man who journeyed from the Earth to the Moon, and exchanged ideas with the Moon-men, or Selenites (from Selene, the classical Greek goddess of the Moon).

Then, a few years later, Francis Godwin – a young English clergyman who was destined to become Bishop of Hereford – wrote his *Man in the Moone* (1638). This was not only another piece of space-travel science fiction, but contained a description of the hero's ascent into space, from where the green continents became smaller, the sea bluer, and the black sky gradually filled with stars – which would not be seen in reality until the 1960s.

But the most far-seeing and ingenious space-travel thinker of this age was the Reverend Dr John Wilkins (1614–1672), Warden of Wadham College, Oxford, and later Bishop of Chester. Wilkins embodied that combination of religious commitment, scientific passion,and optimism for the possibilities of the new science which epitomised the most ingenious intellects of that age. His *The Discovery of a New World... in the Moon... with a Discourse Concerning the Possibility of a Passage Thither* (1638 and 1640) and his *Mathematical Magick* (1648) brought the consequences of Galileo's telescopic discoveries and the new instruments and machines of the age before a wider, English-speaking audience; for among other things, Wilkins was an early populariser of science.

In his writings, Wilkins discussed ways in which one might fly into space, and among several options he decided upon the Flying Chariot, or some kind of spring-driven, flapping-winged vehicle, as the best way of ascending above the clouds. From his reading of Gilbert's *De Magnete*, and of Kepler's and other scientists' works, along with some experiments of his own, Wilkins concluded that the Earth's own 'pull' extended only 20 miles above the surface. If his Flying Chariot, therefore, could rise up this distance, it would escape from the Earth, and fly off into gravity-free space, from whence the aerial navigator could lay his course for the Moon. And just as with Bacon and most of the other inspired men of that age, it was to Columbus, Magellan and Drake that they looked for role models as explorers; for just as they had broken the bounds of old Europe to find continents anew, so Wilkins stimulated his putative travellers to explore whole *worlds* anew.

Although we now realise that Wilkins' writings contained more inspiration and wishful thinking than it did practical science, his flying machines and suggested preparations for a neo-Columbian passage to the Moon from Cromwellian England brought together the best scientific knowledge that was available in the 1640s. And Wilkins' standing as an experimentalist and as an inspirer and organiser of science is beyond doubt. This admirer of Bacon, Galileo, Tycho, Kepler and Gilbert was no mere dreamer, but actually built experimental flying machines and tried them out in the gardens of Wadham College, as his protégé, Dr Robert Hooke, testified many years later in the 1670s. In Oxford and in London, moreover, the eminently clubbable and obviously charismatic Wilkins drew around himself a body of scientific friends – anatomists, astronomers, chemists, physicists and mechanical philosophers – who formed England's first scientific club. And when, after the Civil Wars, the Stuart monarchy was restored in 1660, that very scientifically-minded monarch, King Charles II, gave official incorporation to Wilkins' Club, after which it became the Royal Society of London. Three and a half centuries later, the Royal Society is still an international benchmark for excellence in scientific research, and is Great Britain's oldest and most distinguished learned society.

By the time of his death in 1672 – in the full dignity of the Bishopric of Chester, to which King Charles had elevated him – Dr Wilkins had realised that a journey to the Moon was impracticable. A host of new discoveries made in England and across Europe over the previous 30-odd years had quite undermined many of the scientific assumptions upon which his proposals of 1640 had been based. But that in itself was a measure of the speed with which the new scientific movement was developing by this time.

Before 1640, for instance, scientists knew nothing about the existence of the vacuum which, in a divinely complete and full Creation, was generally believed could not exist, for its presence would indicate some kind of gap. However, in 1644, in Italy, Evangelista Torricelli discovered an evacuated space sealed in above the mercury column of his newly invented barometer when the air pressure fell; and by 1658, further investigations in Italy, France, Germany and England proved beyond doubt that vacuums really did exist. Then, in 1658 Robert Boyle and Robert Hooke, using the world's first 'modern' vacuum pump in a private laboratory in Oxford, discovered that nothing could breathe or sustain itself in a vacuum. This knowledge was corroborated by the results obtained by French scientists who had carried a barometer up a mountain in the Pyrenees, and found that as they became progressively breathless with altitude, the mercury level in the barometer fell. Could it be that the Earth's atmosphere became *thinner* with altitude, and that astronomical space itself was a vacuum? If this were the case (as indeed it is), Wilkins' machine could never fly in space because there would be no resisting medium against which its wings could beat. Neither would there be anything for his space-voyagers to breathe!

We have already seen how, in the wake of the demise of the crystalline spheres and of Aristotle's animistic physics (where heavy things fall to be with 'mother Earth'), astronomers were at a loss to be able to explain why bodies actually moved, and magnetism was one of the tentative explanations put forward. By the 1650s, however, it had come to be realised that the force which makes a stone fall is not the same as that which moves the needle of a magnetic compass. Then, in 1662 Robert Hooke began to perform a series of experiments in the high vaulted galleries of Old St Paul's Cathedral (the cathedral which burned down in the Great Fire of London in 1666) to determine whether an object weighed slightly less when placed 90 feet above the ground than it did at ground level. Hooke's mind was clearly working along the radically original line of seeing weight as *relative* to location, and not as an absolute. What Hooke was attempting in these experiments was to find a mathematical or proportionate expression for changes in weight with respect to location that might indicate the existence of a 'gravitating principle'.

Robert Hooke was unable to draw any firm conclusion from these and similar experiments – not because his basic principles of enquiry were wrong, but because the instruments that were available in 1663 were not sufficiently sensitive. Even so, a succession of scientists – including Descartes, Gassendi, Horrocks, Hooke and Christiaan Huygens – had begun to search for a unifying force that crossed space, controlled the motions of the planets, and even produced the elegantly curved orbits in which cannon balls flew. Most of these men were thinking along the lines of there being some kind of *mechanical* force at work, that explained both terrestrial and astronomical physics through the same set of mathematical laws.

The comprehensive answer was found, of course, in the Laws of Universal Gravitation formulated and defined by Sir Isaac Newton in 1687, although Newton broke with the mechanical tradition and with Descartes' imaginary swirls and vortices of corpuscles. Indeed, Newton abandoned all attempts to explain planetary motion in terms of one thing pushing another, and argued instead that while he did not know the nature of gravity in absolute physical terms, he could, nonetheless, calculate exactly how it operated between bodies, depending upon their respective masses and distances apart. In his abandonment of assumed mechanical causes in favour of exact mathematical descriptions of how bodies move in space, Newton completed that search for the true cause of planetary motion which had begun with Tycho's and Kepler's realisation that the ancient crystalline spheres did not, in fact, exist.

While Bishop Wilkins' hopes for new voyages of discovery across the solar system had clearly been dashed by the new knowledge of the mid-seventeenth century, the possibilities of what might exist in an infinite Universe had grown in fascination. Many of these possibilities took on unexpected religious connotations. If, for instance, the Sun was but a star, then why should other stars not have planets rotating around them? And if this was the case, could these worlds be inhabited by intelligent creatures like ouselves? Indeed, we often think of the idea of life on other worlds as very modern, but in reality its serious discussion extends back to the seventeenth century.

What also interested John Wilkins and many others was the spiritual status of such extraterrestrial beings. Could they have immortal souls like humans; and if so, were they saved or damned? For surely, if the beings that lived on planets rotating around Arcturus or Sirius had never committed original sin, then they could not be a fallen race like earthly humanity, and must therefore be spiritually superior to us. Alternatively, if they were indeed sinners like us, did it mean that God would have to become incarnate in Jesus Christ and be crucified on *every* inhabited planet in the Universe, as a way of redeeming all of these extraterrestrial people? Either way, mankind's unique spiritual status and special relationship with God seemed somehow compromised. No-one had a clear answer to these questions about the 'plurality of worlds', but what the very fact of their being asked shows us is how astronomy and religious belief were undergoing a constant and dynamic interchange, especially in the wake of new telescopic discoveries after 1610.

Far from undermining religious belief, however, the new knowledge in astronomy and the other sciences which became available in the seventeenth century tended to strengthen it. For while this knowledge undoubtedly produced awkward paradoxes – such as that concerning the spiritual status of possible extraterrestrials – all of this new science had two powerful influences on religious understanding. Firstly, it emphasised the design principle present in nature; for no matter where a scientist looked with his new instruments of investigation – be it among the living *cells* which Robert Hooke first saw and named when using his microscope, or at the mysterious formations within the Orion Nebula observed by Christiaan Huygens with a powerful new telescope – orderly structures abounded in nature. And secondly, all of these discoveries served only to reinforce the unique and special character of human intelligence which, far from being worn out and inferior, clearly possessed an astonishing potential for further development.

When these two new realisations – a constant design running throughout nature, and the flexibility of the human mind to comprehend that design – were combined, a wonderful and far-reaching vision became possible. It seemed that Kepler's vision of the astronomer 'thinking God's thoughts after Him', and of the disciplined human intellect following the beautiful warp and weft that the Almighty had woven through the very fabric of the cosmos, now lay spread out before humanity. Scientists like John Wilkins, Robert Boyle, Blaise Pascal, Isaac Newton, and many others, were enchanted by the obvious symbiosis which God had clearly intended to exist between the structures of atoms and star systems on the one hand, and the human mind on the other. They saw it very much in terms of the divine Book of Nature of which Galileo had spoken to the Grand Duchess Christina in 1615, which complemented and perhaps even amplified the older Book of the Word, or Scripture. Nature, of course, did not amplify Scripture in the respect that God was saying something through science that he had not already revealed to the ancient Jews and early Christians in the Biblical texts, but only inasmuch as he was now going into more detail. Moses and Elijah did not need to know that the Earth went around the Sun when delivering God's laws and prophecies to the Jews, any more than the witnesses to the crucifixion needed to know that the wooden cross to which Jesus was nailed was itself made up of microscopic spherical cells arranged in exact geometrical formations.

Science, therefore, was seen as revealing further details of God's glory as a form of continuing intellectual revelation. It is hardly surprising, then, that these scientists felt both *privileged* and *humbled* as they went about their work, and that Boyle offered up prayers to God from his laboratory when each new aspect of nature was revealed to him in the form of a scientific discovery.

The scientists of the seventeenth century – especially those in England – came increasingly to speak of a *Natural Theology*, whereby nature's design, in combination with human scientific reason, was seen as carrying a powerful message about the divine. Its greatest early exponent was the chemist Robert Boyle (the discoverer of Boyle's Law of gases), a deeply devout layman and private gentleman whom his friend and biographer, John Aubrey, styled almost a 'Lay-Bishop'. Indeed, had he agreed to be ordained, Boyle would almost certainly have become a bishop; but he refused all public offices and oaths, saying that he preferred to defend the Christian faith from conviction rather than from legal obligation. Yet while he was primarily a laboratory scientist, Boyle's most vivid analogies of the divine design of nature were astronomical. Like Richard of Wallingford and his fourteenth-century forebears, Boyle was aware of the parallel that existed between the mind of God as reflected in the Universe and the mind of man as shown in our attempts to replicate the celestial movements in clockwork. For, said Boyle, was not the Universe like that extraordinary piece of astronomical automata, the Great Clock of Strassburg? And like that clock, must not the Universe have a rational designer?

It would be incorrect to undervalue the scientific achievements of European scientists in the seventeenth century; but for a variety of reasons deriving, perhaps, from the growth of economic prosperity, the beginnings of libertarian government, and a state Church of England which was both intellectually tolerant and contained within its ranks some of the finest minds of the age, England found her Renaissance voice in science.

TWELVE

Astronomy, religion and culture from 1650 to the present day

It would be too simple to consider seventeenth-century English natural theology as providing the definitive statement on matters of science and religion, for natural theology contained within itself many potential problems. While Boyle might speak of the Universe as being a greater version of the Great Clock of Strassburg in terms of its design, one might well counter this argument by pointing out that even in Boyle's time the Clock's builder had long been dead, though his handiwork continued to mark the hours. What proof existed, therefore, that the God who had, at some stage in the past, built the Universe, was still listening?

The belief in a designing God who had withdrawn from his Creation, or even died, came to be called *deism*. It came to be especially prevalent in continental Europe, and most of all in the salon society of eighteenth-century Paris among those *philosophes* who could not, in that increasingly scientific age, deny that the Universe possessed a coherent structure, and yet could not believe in a transcendent God.

The point at which so many deists parted company from traditional Christians was on the subject of miracles. Quite simply, how could God intervene – even if he wanted to – in a Universe which was coming increasingly to be seen as bounded by exact laws which governed everything from the movements of atoms to the stars of the Milky Way? Was he not excluded by the sheer perfection of his celestial handiwork?

It seemed, therefore, that one was forced to choose between two equally unacceptable images of God, especially in this post-Newtonian, gravitational physics-bounded age. Either one believed that God was – in accordance with the Scriptures – an omnipotent and all-powerful Creator who had made the Universe so perfectly that His further intervention was unnecessary; or else He was the equally scriptural loving father who was obliged to continue meddling with His Creation because he had failed to 'get it right first time'. This latter view of the Creation came to be styled by David Hume in his *Dialogues concerning Natural Religion* (Part V (1779)) as the 'first rude essay of some infant deity, who afterwards abandoned it, ashamed of his lame performance' – a botching and bungling Creator who, well intended as He might be, had to keep responding to His children's prayers to save them from shipwreck or famine because He had not foreseen what was going to happen in advance, and thereby built the necessary correcting forces into His original design.

Yet this stark choice was not the only way forward, for, as figures such as Boyle, Wilkins, John Ray, Bishop Butler and many others argued, one could also see God as interacting with the world for humanity's sake in very much the same way as a wise parent intervenes on behalf of children whom that parent recognises as moral agents in their own right. And this did not mean that God had failed to foresee what would happen, but that we were being allowed to use that free will that He had given us – even if we sometimes used it foolishly and then cried for help.

This way of thinking placed less emphasis on the inevitable determinism of nature, and more on what came to be seen as *Providence*, or the essentially favourable disposition of nature to mankind. For agriculture, navigation, medicine and the arts of civilisation were all seen as part of this broad Providence, and all of them extended back to the dawn of recorded history. And as time went on, the human race was gently steered to yet greater knowledge and capacity by God's guiding hand. It is true that people still died horribly and apparently undeservedly, but the developers of this science-based natural theology during the eighteenth century pointed to the general tendency of things to improve under the influence of this divine hand; and as Alexander Pope succinctly expressed it in his *An Essay on Man* (1734): 'The Universal Cause, Acts not by partial, but by general laws'. For while sin could be the cause of all manner of mischief, God's design was clearly beneficent. This attitude was nowhere more aptly summed up than in William Paley's *Natural Theology* (1802), in which he argued, amongst other things, that 'teeth are made to eat and not to ache'. Or, to express it differently, the divine design was good, though human sin, neglect and folly could cause it to go wrong. This view of Providence, therefore, tended to be excellent when explaining the 'big picture', but left much to be desired when dealing with individual circumstances.

Indeed, many of these Natural Theological writers, in their delight at the seeming superabundance and providence of nature, often failed to face those ancestral problems not only of Christianity, but of all the world's great religions that address the broader human condition. Why, for instance, do innocent children die; and why, as the *Psalms* never failed to remind God, do good people languish, while sinners prosper? Why are God's laws so good in their wider or general design, yet so apparently careless when dealing with special or individual cases?

Indeed, the answers to these questions have nothing to do with science and cosmology, but relate to broader issues of moral purpose and ethics. In many ways, one of the disservices which optimistic natural theologians did for religion, and for Christianity in particular, was to seemingly duck these wider religious concerns in their fascination with the beauty of cosmological perfection. For what these men did – in their belief that God's goodness could be somehow demonstrated on the strength of scientific *evidence* – was to back religion into a corner, for natural theology was powerless to address the eternal problems of individual pain and suffering.

One response of intelligent people to this situation was to split religious and scientific thinking apart from each other, and in some quarters science began to develop as a self-contained culture that either paid conventional lip-service to religion, or else completely ignored it. The Positivism of Auguste Comte, for instance, which grew up in early nineteenth-century France, discounted all knowledge that was not susceptible to scientific, or *positive*, demonstration, and saw transcendental religion such as Christianity as the shrinking remnant of a superstitious past that was destined to vanish away. Recognising the importance of ritual, however, he proposed a secularist religion of humanity. Comte's *Cours de Philosophie Positive* (1830–42), moreover, was clearly influenced by the self-acting and self-sustaining cosmology of the great French mathematician Pierre Simon Laplace, who is alleged to have once told Napoleon Bonaparte that he had no need for the God 'hypothesis'. Positivism won support outside France, and in 1853 Harriet Martineau translated his works into English under the title *The Philosophy of Comte*. Likewise, during the 1850s the Positivist-influenced Herbert Spencer developed a radical secularist philosophy in which the forces of progress in social evolution grew out of conflict and obsolescence, in what Spencer (and *not* Charles Darwin) first styled as 'the survival of the fittest' in his *Principles of Biology* (1864–67). Darwin liked the phrase, however, and incorporated it – with full acknowledgement to Spencer – in the fifth edition of *On the Origin of Species* in 1869.

Yet if the rejection of religion was one response to the problems posed by scientific ideas, then the effective *ignoring* of science and a total concentration upon humanity's spiritual predicament was the other response. Methodism – which began with John Wesley in the eighteenth century, and which grew enormously in the nineteenth century – placed its whole concern on the moral regeneration of individual sinners, through the necessary process of personal repentance. And at the apparent opposite end of the ecclesiastical spectrum from Methodism, the rediscovery of the English Catholic tradition, as epitomised in the works of John Henry Newman and John Keble, similarly emphasised the need for a new and more profound spiritual awareness, based in this case on the recovery of that historical Christianity supposedly lost during the Protestant Reformation. Between these seemingly spiritual extremes were theologians and lay men and women who believed that science had revealed divine wonders, and yet who saw no fundamental conflict with traditional teaching. This group, indeed, contained some of the most influential intellectuals of the Victorian age: astronomers such as Sir John Herschel, the Revd Charles Pritchard and Agnes Clerke; academic physicians such as Sir Henry Acland, and reforming statesmen such as William Ewart Gladstone.

While it is true that Darwin's *On the Origin of Species* undermined the straightforward argument from design that Paley had put into its definitive form in 1802, late nineteenth-century theologians and devout scientists were less concerned with proving the existence of God through the complex beauty of nature. Instead, they thought of that beauty and scientific understanding as a form of contemplative aid, whereby the mind was uplifted by reason to cross the bridge of faith, to a suprarational sense of the total glory of God. This was, after all, the God of historical Christianity, whom in many ways the Victorians rediscovered after the Protestant Church's flirtation with attempts to prove divine truths from scientific evidences had been found to be going down a blind alley. This God to whom Wesley,

Newman and others drew attention, was shown to be far more than just a clever Cosmological Mechanician, but also a Redeemer and Comforter of the World, whose frail human creatures might well perish in time, yet still rise again to live in eternity.

We have seen (in Chapter 1) how Christian intellectuals came to terms with Darwinian evolution by recognising that while our bodies may have arisen from earlier organic prototypes extending across millions of years, our souls, as epitomised in our sense of the divine, were nonetheless quite separate from and transcended this physical creation. Likewise, thinking Christians responded positively to the rapid discoveries in late nineteenth-century astronomy. The new science of astrophysics – which also grew up in the 1860s – showed that the same chemical elements that made up our bodies were also present in the Sun and stars, and that a remarkable uniformity of chemical composition and design seemed to run through the Universe. And that same science of astrophysics, when combined with the development of new, powerful photographic telescopes, was revealing the Universe to be unimaginably vaster than Galileo or Wilkins could ever have conceived. Talking about infinity was one thing; but photographing and chemically analysing it with the spectroscope was an entirely different matter.

This increasingly exact knowledge about cosmological infinity destroyed the religious faith of some astronomers – such as Isaac Roberts, and Thomas Hardy's fictitious hero of *Two on a Tower*, Swithin St Cleve – yet it inspired others to a sense of the celestial majesty of God. Angelo Secchi, Stephen Perry, Walter Sidgreaves and Aloysius Laurence Cortie, for example, were all Jesuit priests as well as professional astrophysicists whose Christian faith inspired them to push scientific knowledge to the very edge, confident that whatever they discovered, it was made by the same God in whose image they stood. And no-one can read the published books and private observing journals of the Revd Thomas William Webb without sensing that

whenever he looked through his telescope he felt that he was behold-
ing his Creator's handiwork. While the Universe might be breathtaking in its immensity, and we as its beholders might seem insignif-
icant specks within it, nevertheless we were observing it, and our
God-given gift of reason enabled us to systematically probe its deepest secrets.

By the late nineteenth century, therefore, the earlier shocks deliv-
ered by the Laplacian Positivists and by Darwinian evolution were
coming to be absorbed, and while many people lost their faith, an
equally large, if not larger, body of thinking men and women felt that
it had been strengthened by the ordeals of the mid century.

One of the extraordinary features of the nineteenth century in
Britain, America and Europe was the almost complete relaxation of
official religious conformity. People were no longer being required to
go to Church by law. Europe's Jews were no longer being made to live
in ghettos or wear distinctive clothing, and while Roman Catholics still
tended to be mistrusted in England, the old professional, political and
legal disabilities under which they had practised their faith since the
Reformation were finally lifted in 1829. And if one were willing to
brave the sense of social ostracism that could go along with it, one was,
by 1870, at liberty to believe in any kind of god that one chose, or in
none, without fear of official censure. After 1858, English Jews could
take their seats as Members of Parliament without being required to
swear an oath of allegiance on the New Testament, while by 1874
Oxford and Cambridge Universities had abandoned their old religious
'tests' which for centuries had restricted access to Anglican Christians.

As a result, all sorts of broadly religious innovations began to take
place within the leisured middle classes – from a serious interest in
mainstream Eastern religions such as Buddhism and Hinduism, to
dabbling with the occult and latter-day druidism. Spiritualism, for
instance, came out of America with the Fox sisters in the late 1840s,
and between 1855 and 1870 the Scottish–American medium Daniel

Dunglass Home caused a sensation before the English royal and French imperial families with his spiritualist demonstrations. Indeed, in this world of increasingly fluid religions, spiritualism had a strong appeal to those who passionately wanted to believe in a life after death, yet somehow felt ill at ease with traditional Christianity. In this new age of man's control of nature by means of applied science, moreover, it was comforting to some people that mediums seemed to be able to communicate with the spirit world with no more difficulty than if they were posting a letter. Yet virtually all of the mediums, when scientifically tested, proved to be frauds: all, that is, except Home, whose feats of levitation and apparent ability to play distant musical instruments still cannot be explained today.

Then there was Madame Helena Petrovna Blavatsky, whose Theosophical Society mixed Indian religions with the mysteries of esoteric Tibet and other beliefs. By the end of the nineteenth century an exotic collection of alternative 'religions' was arising, including Druidism, table-rapping spiritualism, occultism, ghost-hunting, yoga, astrology and, by the early twentieth century, Alistair Crowley's 'Golden Dawn', in addition to those which rode more obviously on the back of Christianity or the old Eastern faiths. America on the whole tended to be the home of most of the Christian-related movements, such as Mary Baker Eddy's Christian Science. Then, also from America, there was Mormonism, and the neo-apocalyptic Jehovah's Witnesses and Seventh Day Adventists, with their concern with calculating the end of time and separating the new 'chosen people' from the rest. Britain, on the other hand, with its Indian Empire, tended rather to produce offshoots from authentic Oriental spiritual traditions, such as Theosophy, although in this speeding world of trains, steamships and cheap printing, lines of demarcation quickly blurred, as comfortably-off and invariably eccentric spiritual supermarket shoppers began to assemble their own faiths. What one is seeing here, as incense is burned before statues of the Buddha in chintz-curtained rooms in Bloomsbury, is one of the first glimmerings of the New Age movement. Here one finds spiritual tourists – attending a séance here, listening to a lecture by a specially imported or plainly bogus Indian guru there, playing with ouija boards, and meeting with like-minded friends to sit on the drawing-room carpet and say 'Om.....' before the maid brings in vegetarian sandwiches and herb tea, or whatever the latest 'nature diet' specifies. It was a world, indeed, that one finds discussed and catered for in the magazine literature of the late Victorian and Edwardian periods, and which is playfully sent up in E.F. Benson's *Queen Lucia* (1920) and *Lucia in London* (1927).

But how do the acolytes of this spiritual world – be they those in Edwardian drawing rooms, or the Sun-worshippers found at Stonehenge on midsummer morning – differ from the practitioners of the older religious traditions that supposedly inspire them?

Well, anything that is supposed to continue in the Wicca, Druid or other pagan traditions from the 'olden times' is transparently bogus, for any significant authentic knowledge of these traditions was wiped out in Roman or early medieval times, and what we have nowadays is an *invented* 'tradition', although it is true that recorded fragments about the Druids have come down through largely hostile late Greek and Roman writers such as Strabo and Tacitus. And as we have seen (in Chapter 1), the Druids' association with Stonehenge derives from a casual remark by the seventeenth-century archaeologist John Aubrey, who first surveyed Stonehenge, and who, from his then limited knowledge, suggested that the ancient and semi-mythical Druids may have built the monument.

Druidism became popular during the Romantic Age of the late eighteenth century, when England's original hippies allowed their imaginations to run wild. Under the inspiration of the Druidic eccentricities of the otherwise scholarly archaeologist Dr William Stukeley, and of the image of the white-robed bearded sage, complete with oak leaves and arcane ornaments, monuments such as Stonehenge and

Avebury became blank sheets upon which the romantic imagination was free to paint what it wished. Since that time, and especially in the latter half of the twentieth century, when it came to be augmented with neo-pagan and New Age notions, Druidism has spun yet more webs of historically challenged mythology about itself.

But why, one might validly ask, in a society which no longer enforces religious conformity, is not belief in magic, Druidism, astrology, crystal healing, witchcraft, and all the rest of neo-paganism, perceived to be of equal spiritual value to the great historical religions? Why is not a self-styled white witch in Milton Keynes capable of providing spiritual insight that is just as valid as one might expect to receive from the Pope, the Chief Rabbi, a great Koranic teacher, or a holy man or woman from one of the mature Oriental faiths?

A New Age devotee might argue that in speaking of a 'mature' faith one is displaying some kind of prejudice in favour of the old and the established, though for a neo-pagan to argue thus, he or she must realise that they are handling a two-edged sword. For, when asked, neo-pagans generally describe themselves as practising the *old religion*: some sort of ancestral Earth-cult which originated with the Druids and wizards of yore, and which was somehow destroyed or driven underground by the newer religions of Christianity, Islam, and even Judaism. And yet, if the primal Earth-cult really was our ancient and ancestral faith, now somehow re-realised by today's pagans, why was it so insubstantial as to be effectively wiped out by the new religions? Could it be that what it had to offer was not worth preserving?

Is modern paganism, therefore, the authentic inheritance of a genuinely old and seemingly forgotten religion, which somehow lacked the spiritual and intellectual stamina to survive; or is it an 'immature' collection of nostalgic yarns that lack any real coherence? I believe that this sense of identity crisis runs deeply through the modern pagan movement, and is characterised by the disparate and confused nature of its thinking.

I would also be so bold as to suggest that neo-paganism and other self-assembled 'supermarket' faiths do not have coherence for another reason: they ultimately lack both intellectual and spiritual discipline. For what makes a religion 'mature', and able to weather the storms of the centuries, is its ability to engage with the rational intellect and the human spirit at the deepest levels; and this can be tough and demanding. But neo-paganism is a 'comfy' religion – a product of a consumer culture, that you can have or not have, as and when you want. Yet the great religions that have shaped the human race's spiritual and moral perceptions have never pretended to be soft options, and have never shrunk from putting their devotees through their spiritual paces.

Indeed, one might further argue that on one level a religion is a species of social technology, and that like all technologies, we value it for its usefulness. This usefulness can best be seen in terms of the religion's explanatory and inspirational power.

One of the central arguments which has run through this book is that modern Western scientific culture came into being through recognising the presence of a design principle in nature that was imparted to it by a singular and supreme God, in whose loving and rational nature all human beings were believed to partake. That rational theology, or intellectually justifiable belief in the transcendent, and a logical understanding of the Universe which we call 'scientific', all came, indeed, from this concept of a monotheistic Deity. For I would argue that the Graeco-Judaeo-Christian religious tradition became so formative in the creation of the modern world because it offered more credible solutions to more problems – from explaining the motions of the heavens to the personal immortality of the human soul – than anything which had gone before.

Moreover, all of these expressions of the One God – from Yahweh, Plato's *nous*, Christ's Sermon on the Mount, and the Islamic Sufis – were capable of empowering mankind, by leading us onwards and upwards in a search for a higher truth that would set us free from darkness. They did this by making us recognise the image of God within ourselves, and by teaching us to aspire, in spite of our sins and defects, towards a higher and nobler physical and spiritual condition. In short, they inspired us to search for order and perfection in the heavens, and for morality, coherence and altruism amongst our fellow creatures.

When they spoke of *truth*, moreover, they did so with that confidence of vision which pre-dated post-modernism's self-doubt, and which knew the difference in value between a *pot pourri* of beliefs that were at best a confused personal indulgence, and a religion which could release powerful creative energies; a religion, indeed, which could inspire great literature, build hospitals, feed the poor, create universities, and discover new lands. In the long run, one judges a set of religious beliefs by criteria similar to how one judges an invention, a book, or a political system: by its fruits and by its results. Does it leave you where you started, pull you back, or inspire you to go forward? Does the religion have the expansive power to 'go public', react with the world, and change it for the better; or does it sink into self-obsessed sub-cultures that achieve nothing enduring? And central to this sense of direction, I would suggest, is the religion's capacity to actively partake in rational discourse about where things come from, how they work, and how the divine and the human interrelate within a deliberately created and logical scheme.

Yet also lying at the heart of much New Age thinking is a laudable respect for nature; and while I would contend that it is both spiritually and intellectually inadequate to venerate nature as though it was a divine being in its own right, one is forced to ask why many people in the late twentieth century came to feel so disenchanted by and alienated from *scientific* explanations of nature that they turned instead to more animistic explanations. This, I suspect, has much to do with the cold, merciless, materialist and deterministic interpretations of science that have received such prominence during the last decades of the twentieth century (as mentioned in Chapter 1).

One thing which New Age pagans share with Jews, Christians and Muslims is the belief that there is more to life than just matter; but where they differ is that Jews, Christians and Muslims believe that human beings are special because of their unique, individual immortal souls that have been created by God, as opposed to simply being products of a universal and self-sustaining fertility principle. But scientists and others who are materialists do not believe either of these non-physical options, the lamentable consequence of which is that scientific and religious explanations can be in danger of going their own way, talking only in their own language to their own kind, and failing to interact with each other.

On the other hand, when one looks beyond the dogmas of the 'fundamentalist materialists', one finds all manner of new and exciting possibilities emerging, and ways in which science and theistic religion – and especially Christianity – are beginning to relate to each other. And while no-one is talking of the Argument from Design as it was articulated by William Paley two centuries ago, modern theologians and scientists are generally willing to argue that both the Universe and the human mind which perceives it are *wonderfully orderly structures*.

What is more, modern cosmology is full of possibilities for potential theistic or God-directed interpretations, several of which seem strangely reminiscent of Thomas Bradwardine and Nicholas of Cusa. Gone is the Newtonian Universe of one dimension of time and of space. Instead, post-Einsteinian cosmology and post-quantum physics have opened our minds to all sorts of peculiar states of being. We now know, for instance, that space is curved, that time is relative to one's velocity with relation to other objects and to the speed of light, and that energy and matter are the same thing. All of these things come together, moreover, in those objects called black holes, which warp the space and time frames around them, create gravitational forces that are so great that not even light can escape from them, and do strange things with matter and energy.

As the Universe is now understood to be a place of variable time scales, dimensions, and states of being, through which hitherto unimaginable states of energy pass, is it that outlandish to suggest that a divine intelligence might also pass through it, or that heaven itself exists in another dimension of time? When, after about 1950, most astronomers came to agree that the Universe appears to be expanding in all directions and dimensions from some primary Big Bang, this posed the obvious question about what preceded the Big Bang. Suddenly this was no longer the familiar timeless Universe of the pre-scholastic Aristotle or the post-seventeenth-century Newtonian Universe of infinite stability, but a place which had a beginning and, one might reasonably suppose, will have an end. And if it had a beginning, and began with a 'bang', was it beyond the bounds of possibility to think that a Creator might have been responsible for laying and firing the charge?

When one uses the physical data – largely based on what are called the spectroscopic redshifts of distant galaxies – that have been painstakingly collected by observatories around the world to calculate an age for the Universe, moreover, one finds oneself faced with a question that had been familiar to St Augustine. For whether the Universe is a mere 6,000 years old – as Julius Africanus thought – or a breathtaking 10–20 billion years old – as we think today – one still feels obliged to ask why it did not come into being sooner or later than it did, and what existed before it. We now know that when the Big Bang took place, only the lightest elements such as helium and hydrogen existed within the infant Universe, with the heavier elements, such as carbon and iron, forming later. But does this mean that *nothing whatsoever* existed before the Big Bang, and that by definition, neither time nor space existed, because (as even St Augustine had realised) space and time are defined by the objects that occupy it?

No-one, of course, is suggesting that modern cosmological models of warped space, the Big Bang, and a possible prior state of total nothingness, prove the existence of God; yet what they undoubtedly do is to pose profound questions about beginnings, endings, and even purposes. And try as one might, these questions can never be convincingly stripped of theological implications. Whether one argues that these questions arise because of the existence of a genuine divine master plan, or simply because human beings are order-seeking creatures, the Creator possibility will not go away easily.

Twentieth-century scientific research on every level – from particle physics through the whole of the life sciences and to cosmology – has also highlighted another important factor in the scientific and theological debate: aesthetics. Why does the whole of nature appear to possess such coherence and beauty, and why has beauty as an ideal exerted such a fascination on humanity, moulding everything from the way in which we organise our emotional responses to each other, to how we define a work of art?

Aesthetics, moreover, also lie at the heart of how we frame our scientific theories. The Greeks, as we have seen, were unequivocal about the importance of aesthetic considerations when framing astronomical explanations, while few modern scientists would be happy with a theoretical explanation for a piece of natural phenomenon which lacked numerical or intellectual symmetry. So one may ask: 'Why?' Yet even if one concedes that humanity instinctively searches for satisfying aesthetics simply because our brains are biologically programmed to search for order as part of an evolutionary survival strategy, one still cannot ignore the persistent, lurking question: 'Why?'

Once again, while no-one is proposing that we can formally prove the existence of God from our innate aesthetic sense, it nonetheless opens up avenues for discussion on these matters, and makes it difficult to exclude the *possibility* of there being a supraphysical and transcendent dimension to nature.

In spite of the open hostility shown to religion by certain contemporary scientists, the state of dialogue between the scientific and theological communities had become more active and open to mutual exploration by the end of the second Christian millennium than it has been at any time since the Georgian and early Victorian periods. This dialogue had come about primarily as a result of the growing number of theologically charged questions that twentieth-century scientific research had posed in such fields as genetics, neurology, and perhaps most of all, cosmology. Indeed, many universities now have professors and lecturers who teach and conduct research into aspects of the relationship between religion and science, and there are major academic journals and international conferences devoted to these studies.

There are also several major learned societies dealing with science and religion: for example, the Society of Ordained Scientists, founded in 1987, the members of which all possess high-level academic qualifications in science, and are also ordained Christian priests. Some of these men and women remain in full-time employment in scientific and medical research, and yet officiate as priests in their various churches, often in the non-stipendiary ministry. Others have left scientific employment and have become full-time clergy, while still remaining close to the scientific community. Some even hold cathedral appointments and headships of theological colleges. And for scientists who practise their faith without joining the priesthood there is Christians in Science, which works largely with young scientists and science students; while the University of California at Berkeley has the Center for Theology and the Natural Sciences. Moreover, the Science and Religion Forum, founded in England in 1975, includes Christians, Buddhists, Jews, Muslims, and even open-minded atheists.

What they all share, however, is a firm belief that if, as Galileo argued in 1615, the same God that inspired the Bible also made the Universe, then science and religion must be ultimately speaking about the same things. Possessing as we do the God-given gift of rational intellect, it is for human beings to address themselves to the exploration and solution of that realm of ideas, moral dilemmas and physical facts that our Creator has placed before us; for the ultimate concern of both science and religion is the exploration of truth.

Nor is this interest in the relationship between science and religion by any means confined to Christians. Many Jewish and Muslim scholars are also involved in understanding the connection that exists between the Creator and the Creation, as part of an historical quest that stretches back through the centuries. As we have seen, Judaism and Islam spring from belief in the same one God as does Christianity, and Islamic and Jewish thinkers are faced with the same intellectual problems regarding the role, power and spiritual place of science in the modern world.

No-one today, however, is talking about *proving* the existence of God by some kind of logical or scientific procedure. In the past – and most damagingly in the circumstances surrounding the Darwinian controversy in the mid-nineteenth century – this attempt to find proofs for God's existence often hinged on 'God of the Gaps' arguments, in which the Deity was invoked as a way of finding a neat and simple solution to something that was then inexplicable in contemporary scientific terms – such as the creation of the human eye or the reconciliation of irregularities in the motions of the planets. However, as knowledge advanced, and as scientific answers to these and other questions became available, it seemed that God was being progressively forced out of the 'gaps' in scientific discoveries, so that religious explanations appeared increasingly redundant and new Positivistic explanations seemed more relevant. This was, after all, the origin of that Victorian 'crisis of faith' aptly described in Matthew Arnold's poem *Dover Beach* (1867), which spoke of the sea of faith as ebbing out. And while prominent Christian intellectuals such as Asa Gray, Aubrey Moore and Frederick Temple quickly saw the ease with which 'God of the Gaps' arguments could be abandoned in favour of much more dynamic theistic visions of God in the light of contemporary scientific discoveries, many late Victorian people felt that their more simple faith had been badly shaken. As we have seen (in Chapter 1), atheistic and agnostic scientists such as John Tyndall and Thomas Henry Huxley were quick to take up the offensive against the Church as evangelists of the new secularism. During much of the twentieth century, moreover, their intellectual descendants enjoyed great influence, especially when complemented by the writings, lectures and radio broadcasts of the eminent atheist philosopher Bertrand Russell.

As a result of these developments, science and religion had relatively little to say to each other for much of the twentieth century. Scientists tended to do their science (and perhaps flirt with Marxism), whereas religious believers – mainly Christians and Jews – generally practised their faith in a traditional context that preferred to ignore the arguments of figures like Bertrand Russell.

It was only really when those triumphalist science-based technologies and philosophies that first produced Nazi Germany and then the horrors of atomic devastation in Japan made their impact on broader culture that unavoidable religion-related questions about science began to surface. Was science a morally neutral and unstoppable engine with a will of its own, or were scientists obliged to leave their laboratory–temples and engage with the real world of right and wrong? As the medical sciences began their headlong progress after World War II, hitherto marginal or purely theoretical ethical questions entered mainstream debate: such topics as the morality of euthanasia, the artificial life-support of the long-term comatose, genetic cloning and engineering, and doctors 'playing God' when deciding which patients would receive the new life-saving therapies. All of these, and more besides, became topics on which the developers of complex scientific procedures found themselves facing essentially religious dilemmas, insofar as they bore directly upon questions about the beginnings, purposes and endings of human lives. And in addition to the ethical implications of the life sciences, we had the questions about the 'means and ends' of creation implicit in the new physics and cosmology (as we have seen above).

Long before the 1960s and 1970s, therefore, scientific and religious thinkers were no longer dealing with the 'God of the Gaps' issues; nor were they trying to find Paley-type 'evidences' that would provide simple physical proofs of God's existence and beneficence. With the exception of fringe groups in Bible-belt America, they were certainly not concerned with disproving evolution or proving the literal truth of *Genesis*, and one of the palpable absurdities of certain prominent 'fundamentalist materialists' of the present day is (as we saw in Chapter 1) their continual and perhaps deliberate missing of this crucial point, and their pretence that all religious people are Bible-punching Darwinophobes.

In this dialogue between science and religion, moreover, the Roman Catholic Church has also played a leading role. While Galileo did not have such a felicitous relationship with the Jesuit Order as one might have wished, the Society of Jesus nonetheless provided the Catholic Church with some of its greatest astronomers and scientists throughout the ensuing 370 years. As we have seen (in Chapter 10), the Church began its gradual rehabilitation of Galileo in 1822, with Galileo's *Dialogue* and other Copernican books being removed from the Index of Prohibited Books in 1835. In 1942 the case against Galileo was further examined by the Pontifical Academy of Sciences as part of the scholarly attention aroused by the 300th anniversary of Galileo's death, and early on in his reign in 1979, Pope John Paul II ensured that the Galileo affair was re-examined by the Church. In 1983, His Holiness apologised for the Church's error in condemning such an impeccably devout Catholic in the first place, and by 1992 Galileo had been fully exonerated. In the Pontifical Academy of Sciences (the Roman Catholic Church's ecumenical scientific academy) housed within the Vatican City, one is constantly reminded of the greatness of Galileo and of the ancestral respect in which he is now held by the present-day Academicians and Catholic scientists. A similar respect pervades the Vatican Observatory at Castelgandolfo, near Rome, where the present-day astronomers, with their telescopes, occupy the upper floors of the same palace which acts as the Pope's summer residence.

Indeed, Pope John Paul II took an active role in bridging the gap that still existed between religion and science, and in promoting dialogue. In 1987, moreover, His Holiness gave succinct expression in a letter to Father George Coyne, S.J., Director of the Vatican Observatory, to how the Church now views science, and how each can grow stronger from an awareness of the insights which the other is capable of giving: 'Science can purify religion from error and superstition; religion can purify science from idolatry and false absolutes. Each can draw the other into a wider world, a world in which both can flourish.' (Published in Maffeo, *In the Service of Nine Popes*, p.223.)

Mankind's understanding of the natural world, and especially of the heavens, has come a long way since the Egyptians of 3500 BC first noticed that the pre-dawn rising of Sirius heralded the flooding of the river Nile. And since then, one of the grandest themes in the history of civilisation has been the way in which men and women have realised that nature is not erratic, but can be modelled, explained, and even manipulated, through reason and through the exercise of disciplined intelligence.

Throughout this book we have seen how human intelligence first wrestled with chaos, inventing mythologies to explain the outcome of the nightly battles fought between the Sun and the powers of darkness, or how the world came into being. Yet while the Egyptians, Babylonians, Canaanites and other Near-Eastern cultures produced useful technologies for controlling the flow of water or even predicting eclipses, they never developed a body of *scientific* knowledge, primarily because those planets which moved across the heavens were 'gods in the sky', and not physical objects that were subject to mathematical laws.

It has been my argument throughout this book that Western science, as an impartial study of physical nature, only became possible when people ceased to view astronomical bodies as deities in their own right, and saw them instead as created objects moving in accordance with a plan that their Creator had laid down for them. Without the concept of subject luminaries that were entirely obedient to a Grand Design, there could never have been that coherence across the board, nor that predictability, which is necessary to a rational understanding of nature. In the long run, two quite separate cultural developments were essential for this: the Greek concepts of the *logos* and the *nous*, and the Jewish One God and Creator *ex nihilo*. The first supplied the Western world with the idea of a higher intelligence that embodied within itself perfect prototypes, or *Forms*, that could be imperfectly realised in those physical objects which constituted the natural world. The second spoke not of an impersonal *logos*, but of a personal God – Yahweh – who first created the cosmos out of nothing, and then created the human race in His own image. And both of them replaced the ancient pagan polytheism with a single, unitary monotheism.

When these two ideas – Greek and Jewish – fused together, as expressed in the writings of Philo and in early Christianity, there arose an extraordinary potential for the growth of science, as the impersonal and eternal Greek *logos* and *nous*, with all of their perceived mathematical perfection, were combined with Him who had created all things from nothing, including human beings! And as monotheistic Islam also grew directly out of this early Judaeo-Christian root, and likewise inherited the books and intellectual traditions of the Greeks, one can understand how that rational, scientific investigation of nature which we recognise today began to mature, as it did during the medieval centuries in Baghdad, Damascus, Cairo, Toledo, Paris, Bologna and Oxford.

It was in medieval Europe after about 1150, however, that the basic ingredients for an enduring and astonishingly fertile scientific tradition would come together, and it is a fundamental distortion of well-documented historical evidence to pretend that there was no science worth the name in medieval Europe, and that that which existed was somehow suppressed by the Church. For as we have seen (in Chapter 8), medieval Europe not only created the experimental science of optics, and initiated surprisingly modern-sounding discussions in cosmology, but also began the systematic teaching of astronomy to university undergraduates, and in Sacrobosco's *De Sphaera Mundi* and Chaucer's *Treatise on the Astrolabe*, produced student textbooks in the science. All of this was done with the *full encouragement* of the Christian Church; for without it, there would never have been the physical resources nor the intellectual initiative to follow in the path, first beaten by Cardinal Bessarion and Regiomontanus, that led on to the Scientific Revolution.

That disciplined search for truth which we call science, therefore, is not and never has been a natural foe of religion, and to see it as such is to commit a serious injury to those historical forces which produced Western civilisation. For it was through the growth of belief in an intellectually vibrant, purposive and personal Creator God, through the Greeks, Jews, Muslims, and Christians, that mankind's first perception of an orderly cosmos arose. And it was from that perception that people began to look beyond the 'gods in the sky' to the One God who had made the sky.

GENERAL SOURCES

Astronomy
Crowe, Michael J., *Theories of the World from Antiquity to the Copernican Revolution* (Dover, New York, 1990)
Dreyer, John L.E., *A History of Planetary Systems from Thales to Kepler* (CUP, 1906, reprinted 1953)
Gillespie, C.C. (ed), *Dictionary of Scientific Biography*, 16 vols (Scribners, New York, 1970-80)
Hoskin, Michael (ed), *The Cambridge Illustrated History of Astronomy* (CUP, 1997)
North, John D., *The Fontana History of Astronomy and Cosmology* (Fontana, London, 1994)
Pannekoek, A., *A History of Astronomy* (Allen and Unwin, London, 1961)
Taylor, Eva G.R., *The Haven-Finding Art: a History of Navigation from Odysseus to Captain Cook* (Hollis and Carter, London, 1971)
Thurston, Hugh, *Early Astronomy* (Springer Verlag, Berlin, 1994)
Walker, Christopher (ed), *Astronomy Before the Telescope* (British Museum, London, 1996)

Religion
Holy Bible (numerous editions and translations, eg 1611 Authorised [King James], *New English Bible* (1970), *New International Version* (Hodder and Stoughton, 1989
Bowker, John (ed), *The Oxford Dictionary of World Religions* (OUP, 1997)
Cross, F.L. and Livingstone, E.A., *The Oxford Dictionary of the Christian Church* (OUP, 1997)
Smart, Ninian, *The Atlas of the World's Religions* (OUP, 1999)
Smart, Ninian, *Background to the Long Search* (BBC Books, 1979)

History
Porter, Roy, *The Greatest Benefit to Mankind. A Medical History of Humanity from Antiquity to the Present* (HarperCollins, London, 1997)

ANCIENT CULTURES

Astronomy
Aveni, Anthony F., *Ancient Astronomies* (St. Rémy Press, Montreal, 1993)
Blacker, Carmen and Loewe, Michael, *Ancient Cosmologies* (Allen and Unwin, London, 1975)
Hunger, Herman and Pingree, David, *MUL.APIN.: An Astronomical Compendium in Cuneiform* (Archiv für Orientforschung, Beiheft 24, Horn, Austria, 1989)
Neugebauer, Otto, *Astronomical Cuneiform Texts*, 3 vols (Lund Humphries, London, 1955)
Ruggles, Clive L.N., *Astronomy in Prehistoric Britain and Ireland* (Yale University Press, New Haven and London, 1999)

Religion

Albright, W.F., *Archaeology and the Religion of Israel* (Baltimore, 1966)

Anderson, B.W., *Creation versus Chaos: the Re-Interpretation of Mythical Symbolism in the Bible* (New York, 1967)

Budge, Ernest A. Wallis, *The Ancient Egyptian Heaven and Hell* (London, 1905–25)

Cerny, J., *Ancient Egyptian Religions* (London, 1952)

Cohn, Norman, *Cosmos, Chaos and the World to Come. The Ancient Roots of Apocalyptic Faith* (Yale University Press, New Haven and London, 1993)

Egyptian Mythology (Paul Hamlyn, London, 1965)

Faulkner, Raymond Oliver, *The Ancient Egyptian Book of the Dead* (British Museum, London, 1989)

Faulkner, Raymond Oliver, *The Ancient Egyptian Pyramid Texts* (OUP, 1969)

George, Andrew (trans), *The Epic of Gilgamesh* (Penguin, London, 1999)

Gray, John, The Canaanites (London, 1964)

Grimal, Pierre, (trans Beardsworth, Patricia), *Larousse World Mythology* (Larousse, Chartwell Books, New Jersey, 1976)

Hart, George, *Egyptian Myths* (British Museum, London, 1990)

Heidel, A. (trans), *The Gilgamesh Epic and Old Testament Parallels* (University of Chicago, Phoenix Books, 1949, 1963)

Jacobsen, T., *The Treasures of Darkness: a History of Mesopotamian Religion* (New Haven and London, 1976)

Langdon, Samuel (trans), *Enuma Elish. The Babylonian Epic of Creation restored from the recently recovered tablets of Assur* (Clarendon Press, Oxford, 1923)

Miller, Patrick D., *et al*, *Ancient Israelite Religion. Essays in Honour of Frank Moore Cross* (Fortress Press, Philadelphia, 1987)

Morenz, Siegfried, (trans Keep, Ann E.), *Egyptian Religion* (Methuen, London, 1973)

Quirke, Stephen, *Ancient Egyptian Religion* (British Museum, London, 1992)

Soggin, Jan Alberto, *A History of Israel. From the Beginning to the Bar Kochba Revolt, AD 135* (SCM Press, London, 1984)

History

Breasted, J.H., *Ancient Records of Egypt*, 5 vols (Chicago, 1906-7)

Gardiner, A.H., *Egypt of the Pharaohs* (Oxford, 1974)

Kramer, S.N., *History Begins at Sumer* (London, 1958)

Kramer, S.N., *The Sumerians: their History, Culture and Character* (Chicago, 1963)

Romer, John, *Ancient Lives. The Story of Pharaoh's Tombmakers* (Weidenfeld and Nicholson, London, 1984)

GREEKS AND ROMANS

Astronomy

Aristotle, (trans Stocks, J.L.), *De Caelo* ['On the Heavens'], in *The Works of Aristotle II* (Clarendon Press, Oxford, 1930, 1966)

Aristotle, (trans Ogle, W.), *De Generatione Animalium* ['On the Parts of Animals'], in *The Works of Aristotle V* (Clarendon Press, Oxford, 1912, 1965)

Aristotle, (trans Webster, E.W.), *Meteorologica*, in *The Works of Aristotle III* (Clarendon Press, Oxford, 1931, 1968)

Aristotle, (intro Schoonheim, Pieter L.), *Meteorologica: Aristotle's Meteorology in the Arabico-Latin Tradition: A Critical Edition of the Texts* (Brill, Leiden, 2000)

Aristotle, (trans Hardie, R.P. and Gaye R.K.), *Physics*, in *The Works of Aristotle II* (Clarendon Press, Oxford, 1930, 1966)

Clagett, Marshall, *Greek Science in Antiquity* (Collier–Macmillan, New York, London, 1963)

Dilke, O.A.W., *The Roman Land Surveyors. An Introduction to the Agrimensores* (David and Charles, Newton Abbot, 1971)

Lindberg, David C., *The Beginnings of Western Science* (University of Chicago Press, 1992)

Lucretius [Titus Lucretius Carus], (trans Leonard, W.E.), *De Rerum Natura* ['On the Nature of Things'] (Everyman, London, 1916)

Neugebauer, Otto, *A History of Ancient Mathematical Astronomy*, 3 vols (Springer Verlag, Berlin, 1975)

Pliny [Gaius Plinius Secundus], (trans Healy, John F.), *Natural History. A Selection* (Penguin, London, 1991)

Price, Derek J. de Solla, 'Gears from the Greeks: the Antikythera Mechanism – a calendar computer from ca. 80 BC' from *Transactions of the American Philosophical Society* (NS vol 64, Pt 7, 1974)

Ptolemy, Claudius, (trans Toomer, G.J.), *Ptolemy's Almagest* (Duckworth, London, 1984)

Sambursky, Samuel, (trans Dagut, Merton), *The Physical World of the Greeks* (Routledge and Kegan Paul, London, 1963)

Sambursky, Samuel, *The Physical World of Late Antiquity* (Routledge and Kegan Paul, London, 1962)

Van Helden, Albert, *Measuring the Universe: Cosmic Dimensions from Aristarchus to Halley* (University of Chicago Press, 1985)

Vitruvius Pollio, (trans Granger, Frank), *De Architectura* (Loeb Classical Library, Harvard, 1931, 1998)

Religion

Cicero, Marcus Tullius, (trans McGregor, Horace C.P.), *De Natura Deorum* ['On the Nature of the Gods'] (Penguin, 1972)

Graves, Robert, *The Greek Myths*, 2 vols (Penguin, 1969)

Hesiod, (trans Evelyn-White, Hugh G.), *Homeric Hymns, Epic Cycle, Homerica, Astronomy* (Loeb Classical Library, Harvard, 1914, 2000)

Homer, (trans Rieu, E.V.), *The Iliad* (Penguin, Harmondsworth, 1966)

Homer, (trans Rieu, E.V.), *The Odyssey* (Penguin, Harmondsworth, 1969)

Plato, (trans Grube, Georges M.A.), *Meno* (Hackett, Indianapolis, 1976)

Plato, (trans Gallop, David), *Phaedo* (OUP, 1993)

Plato, (trans Tredennick, Hugh and Tarrant, Harold), *The Last Days of Socrates* (Penguin, London, 1954, 1993)

Plato, (trans Cornford, Francis M.), *Plato's Cosmology, The Timaeus of Plato* (Routledge and Kegan Paul, London, 1937)

Plutarch, (trans Waterfield, Robin), *Essays* (Penguin, London, 1992)

Plutarch, (trans Babbitt, Frank Cole), *Moralia* (Loeb Classical Library, Harvard, 1936, 1999)

Ulansky, D., *The Origins of the Mithraic Mysteries: Cosmology and Salvation in the Ancient World* (OUP, 1989)

Virgil [Publius Vergilius Maro], (trans Jackson Knight, W.F.), *The Aeneid* (Penguin, Harmondsworth, 1966)

History

Aristophanes (trans Henderson, Jeffrey), *The Clouds* (Loeb Classical Library, Harvard, 1998)

Boardman, John, Griffin, Jasper and Murray, Oswyn (eds), *Greece and the Hellenic World* (OUP, 1989)

Boardman, John, Griffin, Jasper and Murray, Oswyn, *The Oxford History of the Classical World* (OUP, 1986)

Cary, M., et al (eds), *The Oxford Classical Dictionary* (OUP, 1949)

Hermann, Paul, *Conquest by Man. The Saga of Early Exploration and Discovery* (Hamish Hamilton, London, 1954)

Herodotus, (trans de Sélincourt, Aubrey), *The Histories* (Penguin, 1968)

Howatson, M.C., *The Oxford Companion to Classical Literature* (OUP, 1989)

McDonald, William A., *The Discovery of Homeric Greece* (Elek Books, London, 1967)

JEWISH, ARABIC AND ORIENTAL

Astronomy

Al-Hassan, Ahmad Y. and Hill, Donald R., *Islamic Technology. An Illustrated History* (CUP, 1986)

Gunther, Robert Theodore, The Astrolabes of the World (1932, Holland Press, London, 1976)

King, David A., *Astronomy in the Service of Islam* (Variorum, Aldershot, 1993)

King, David A., *Islamic Astronomical Instruments* (Variorum, Aldershot, 1987)

King, David A., *Islamic Mathematical Astronomy* (Variorum, Aldershot, 1993)

O'Leary, De Lacy, *How Greek Science Passed to the Arabs* (Routledge and Kegan Paul, London, 1980)

Nasr, Seyyed Hosseinm, *Islamic Science. An Illustrated History* (World of Islam Festival, 1976)

Needham, Joseph, *Science and Civilisation in China*, vol III (CIP, 1959)

Ronan, Colin A., *The Shorter Science and Civilisation in China* (CUP, 1981)

Said, Hakim and Dr A. Zahid, *Al Biruni. His Times, Life and Works* (Hamdard Foundation, Pakistan, 1981)

Saliba, George, *A History of Arabic Astronomy: Planetary Theory during the Golden Age of Islam* (New York University Press, New York, 1994)

Samsó, J., *Islamic Astronomy and Medieval Spain* (Variorum, Aldershot, 1994)

Religion

Al-Qur'an [Holy Koran], A Contemporary Translation, Ahmed Ali (Princeton University Press, New Jersey, 1994)

The Koran, (trans Dawood, N.J.), (Penguin, London, 1956, 1999)

Davidson, Herbert A., *Alfarabi, Avicenna and Averröes on Intellect* (Oxford, 1992)

Donzel, E.J. van, *The Encyclopaedia of Islam* (Brill, Leiden, 1998)

Esposito, John L. (ed), *The Oxford History of Islam* (OUP, 1999)

Guttman, Julius, (trans Silbermann, D.W.), *Philosophies of Judaism: The History of Jewish Philosophy from Biblical Times to Franz Rosenzweig* (New York, 1964)

Houtsma, Matijn T. et al (ed), *Encyclopaedia of Islam*, 4 vols (Leyden, 1913–38)

Husik, Isaac, *A History of Medieval Jewish Philosophy* (1916, New York, 1969)

Philo of Alexandria, (trans Winston, David), *The Contemplative Life, The Giants and*

Selections (Classics of Western Spirituality, SPCK, London, 1981)

Sharif, M.M., *A History of Muslim Philosophy*, 2 vols (Wiesbaden, 1963–66)

Sirat, Colette, *A History of Jewish Philosophy in the Middle Ages* (Harvard University Press, Cambridge, 1962)

History

Norman, Daniel, *The Arabs and Medieval Europe* (London, 1975)

Southern, Richard W., *Western Views of Islam in the Middle Ages* (Harvard University Press, Cambridge, 1962)

MEDIEVAL EUROPE

Astronomy

Chaucer, Geoffrey, *Treatise on the Astrolabe* in *The Works of Geoffrey Chaucer* (Riverside Press, Boston, 1961)

Crombie, Alistair C., *From Augustine to Galileo. Science in the Middle Ages, 5th–13th Century* (Peregrine, 1969)

Grant, Edward (ed), *A Source Book of Medieval Science* (Harvard University Press, Cambridge, 1974)

Grant, Edward, *Planets, Stars and Orbs. The Medieval Cosmos, 1200–1687* (CUP, 1994)

Kuhn, Thomas S., *The Copernican Revolution. Planetray Astronomy and the Development of Western Thought* (Harvard University Press, Cambridge, 1966)

Lindberg, David Charles (ed), *Science in the Middle Ages* (University of Chicago Press, 1978)

Lindberg, David Charles, *Studies in the History of Medieval Optics* (Variorum, London, 1983)

McCluskey, Stephen C., *Astronomies and Cultures in Early Medieval Europe* (CUP, 1998)

North, John David, *Chaucer's Universe* (OUP, 1988)

Turner, Anthony J., *Early Scientific Instruments: Europe 1400–1800* (Sotheby's, London, 1987)

Religion

Augustine, St., of Hippo, *De Civitate Dei* (426) [*The City of God*] (Dent, London, 1945)

Boethius, Anicius Manlius Severinus, (trans Walsh, Patrick G.), *De Consolatione Philosophiae* [*The Consolation of Philosophy*] (OUP, 2000)

Cohn, Norman, *The Pursuit of the Millennium: Revolutionary Millenarians and Mystical Anarchists of the Middle Ages* (Paladin, London, 1970)

Dante Aligheri, (trans Sayers, Dorothy L.), *The Divine Comedy* (Penguin, 1979)

Wegener, G.S., (trans Shenfield, Margaret), *Six Thousand Years of the Bible* (Hodder and Stoughton, London, 1963)

History

Bede, (trans Sherley-Price, Leo), *A History of the English Church and People* (Penguin, 1958)

Boyd, William, *The History of Western Education* (Adam and Charles Black, London, 1954)

Brooke, Christopher, *Europe in the Central Middle Ages, 962–1154* (Longman, 1966)

Brooke, Christopher, *The Twelfth-Century Renaissance* (Thames and Hudson, London, 1969)

Cantor, Norman F., *The Medieval World 300–1300* (Collier–Macmillan, London, 1963)

Cobban, Alfred B., *The Medieval Universities, their Development and Organisation* (London, 1975)

Curtis, S.J.and Boultwood, M.E.A, *A Short History of Educational Ideas* (University Tutorial Press, London, 1966)

Gimpel, Jean, *The Medieval Machine, the Industrial Revolution of the Middle Ages* (Futura, London, 1976)

Heer, Friedrich, (trans Sandheimer, Janet), *The Medieval World, Europe from 1100–1350* (Weidenfeld and Nicholson, London, 1962)

Huizinga, Johan, *The Waning of the Middle Ages* (1924, Penguin, 1968)

Leach, A.F., *The Schools of Medieval England* (Methuen, London, 1915, 1969)

Pacey, Arnold, *The Maze of Ingenuity* (Allen Lane, London, 1974)

Parkes, Henry Bamford, *The Divine Order. Western Culture in the Middle Ages and Renaissance* (Victor Gollancz, London, 1970)

Runciman, Steven, *A History of the Crusades*, 3 vols (Cambridge, 1951–54)

Southern, Richard W., *The Making of the Middle Ages* (Hutchinson, London, 1953)

Sylvester, David W., *Educational Documents 800–1816* (Methuen, London, 1954)

Trevor-Roper, Hugh, *The Rise of Christian Europe* (Thames and Hudson, London, 1965)

Ward, Benedicta, [S.L.G.], *The Venerable Bede* (Geoffrey Chapman, London, 1990)

Wilson, Nigel, *Scholars of Byzantium* (Duckworth, London, 1983)

RENAISSANCE

Astronomy

Allen, Don Cameron, *The Star-Crossed Renaissance: The Quarrel about Astrology and its Influence in England* (Duke University, Durham, 1941)

Broderick, James S.J., *Galileo. The Man, his Work and his Misfortunes* (Geoffrey Chapman, London, 1964)

Casper, Max, (trans Hellman, C.D., Schuman, Abelard), *Kepler* (London, 1959)

Chapman, Allan, *Dividing the Circle. The Development of Critical Angular Measurement in Astronomy*, 1500–1850 (Praxis-Wiley, New York, 1990)

Copernicus, Nicholas, (trans Rosen, Edward), *De Revolutionibus Orbium Coelestium* (Polish Academy of Sciences, Warsaw, 1978)

Copernicus, Nicholas, (trans Duncan, A.M.), *On the Revolutions of the Heavenly Spheres* (David and Charles, Newton Abbot, 1976)

Crane, G.R., *Maps and their Makers. An Introduction to the History of Cartography* (Hutchinson, London, 1968)

Fantoli, Annibale, (trans Coyne, George V.), *Galileo for Copernicanism and the Church* (Vatican Observatory, Rome, 1994)

Fauvel, John, Flood, Raymond and Wilson, Robin, *Oxford Figures. 800 Years of the Mathematical Sciences* (OUP, 2000)

Field, Judith V., *Kepler's Geometrical Cosmology* (University of Chicago Press, 1988)

Galileo Galilei, (trans Stillman, Drake), *Discoveries and Opinions of Galileo* ('Starry Messenger' [1610], 'Letter on Sunspots' [1613], 'Letter to the Grand Duchess Christina' [1615]) (Doubleday, New York, 1957)

Heilbron, John L., *The Sun in the Church. Cathedrals as Solar Observatories* (Harvard University Press, 1999)

James, Jamie, *The Music of the Spheres. Music, Science and the Natural Order of the Universe* (Copernicus, Springer Verlag, New York, 1993)

Kepler, Johannes, (trans Duncan, A.M.), *Mysterium Cosmographicum* [*The Secret of the Universe*] (Abaris Books, New York, 1981)

Koyré, Alexander, *Closed World to Infinite Universe* (Johns Hopkins, Baltimore, 1957, 1968)

Koyré, Alexander, (trans Maddison, R.E.W.), *The Astronomical Revolution* (Cornell University Press, Ithaca, 1973)

Maffeo, Sabino, S.J., (trans Coyne, George V.), *In the Service of Nine Popes. 100 Years of the Vatican Observatory* (Pontifical Academy of Sciences and Vatican Observatory, Rome, 1991)

McIntosh, Gregory C., *The Piri Reis Map of 1513* (University of Georgia Press, Athens)

Parry, John H., *The Age of Reconaissance: Discovery, Exploration and Settlement, 1450–1650* (Weidenfeld and Nicholson, London, 1966)

Thoren, Victor, *The Lord of Uraniborg: A Biography of Tycho Brahe* (CUP, 1990)

Turner, Gerard L'E., *Elizabethan Instrument Makers. The Origins of the London Trade in Precision [Scientific] Instrument Making* (OUP, 2000)

Religion

Bainton, Roland H., *Erasmus of Christendom* (Lion, Tring, 1988)

Bainton, Roland H., *Here I Stand. A Life of Martin Luther* (Lion, Tring, 1987)

Bossy, John, *Christianity in the West 1400–1700* (OUP, 1985)

Dickens, Arthur G., *The English Reformation* (Collins–Fontana, 1964, 1970)

Dickens, Arthur G., *The German Nation and Martin Luther* (Fontana, London, 1976)

Duffy, Eamon, *The Stripping of the Altars. Traditional Religions in England 1400–1580* (Yale University Press, New Haven, 1992)

McGrath, Alister E., *The Intellectual Origins of the European Reformation* (Blackwell, Oxford, 1987)

McGrath, Alister E., *Reformation Thought: An Introduction* (Blackwell, Oxford, 1999)

Thomas, Keith, *Religion and the Decline of Magic: Studies in Popular Belief in Sixteenth and Seventeenth-Century England* (Weidenfeld and Nicholson, London, 1971)

History

Bacon, Francis, *The Advancement of Learning* (London, 1605; OUP, 1966)

Bacon, Francis, *New Atlantis* (London, 1627; OUP, 1966)

Bacon, Francis, *Essays* (London, 1597–1625; OUP, 1966)

Boas Hall, Marie, *The Scientific Renaissance* (Collins, 1962)

Burckhardt, Jacob, *The Civilisation of the Renaissance in Italy* (1860; Phaidon, London, 1944)

Butterfield, Herbert, *The Origins of Modern Science* (Bell and Sons, London, 1957–68)

Davies, C.S.L. and Garnett, Jane (eds), *Wadham College* (Wadham College, Oxford, 1994)

Elton, Geoffrey R., *Renaissance and Reformation 1300–1648* (Collier-Macmillan, 1963)

Koenigsberger, H.G. and Mosse, George L., *Europe in the Sixteenth Century* (Longman, London, 1968)

McLean, Antonia, *Humanism and the Rise of Science in Tudor England* (Heinemann, London, 1972)

Pennington, D.H., *Seventeenth-Century Europe* (Longman, London, 1970)

POST-RENAISSANCE

Astronomy
Westfall, Richard Samuel, *Never at Rest: A Biography of Isaac Newton* (CUP, 1980)

Religion
Atkins, Peter, *Creation Revisited* (Penguin, Harmondsworth, 1994)
Barbour, Ian G., *Issues in Science and Religion* (SCM Press, London, 1966)
Barbour, Ian G., *Science and Religion. New Perspectives on the Dialogue* (SCM Press, London, 1968)
Brooke, John Hedley, *Reconstructing Nature, the Engagement of Science and Religion* (Clark, Edinburgh, 1998)
Brooke, John Hedley, *Science and Religion. Some Historical Perspectives* (CUP, 1991)
Dawkins, Richard, *River Out of Eden* (Weidenfeld and Nicholson, London, 1995)
Dawkins, Richard, *The Blind Watchmaker* (Penguin, Harmondsworth, 1991)
Dawkins, Richard, *The Selfish Gene* (Granada Publishing, London, 1979)
Gay, Peter, Deism, an Anthology (Van Nostrand, Princeton, New Jersey, 1968)
Hastings, Adrian, Mason, Alistair, Pyper, Hugh et al, *The Oxford Companion to Christian Thought. Intellectual, Spiritual and Moral Horizons of Christianity* (OUP, 2000)
Hourani, George (ed), *Essays on Islamic Philosophy and Science* (Albany, New York, 1973)
Hume, David, *Essays* (New Universal Library, Routledge)
Hume, David, *Hume on Religion* (Fontana, London, 1963)
Huxley, Thomas Henry, *Evolution and Ethics* (Pilot Press, London, 1963)
Lindberg, David C. and Numbers, Ronald L. (eds), *God and Nature: Historical Essays on the Encounter Between Christianity and Science* (University of California Press, Berkeley, 1986)
McGrath, Alister E., *The Foundations of Dialogue in Science and Religion* (Blackwell, Malden, 1998)
McManners, John (ed), *The Oxford Illustrated History of Christianity* (OUP, 1990)
Peacocke, Arthur R., *Paths from Science Towards God, the End of All Our Exploring* (One World, Oxford, 2001)
Peacocke, Arthur R., *Theology for a Scientific Age: Being and Becoming – Natural and Divine* (SCM Press, London, 1993)
Polkinghorne, John, *Science and Creation* (SPCK, London, 1988)
Ward, Keith, *God, Chance and Necessity* (One World, Oxford, 1996)
Ward, Keith, *God, Faith and the New Millennium: Christian Belief in an Age of Science* (One World, Oxford, 1998)
Westfall, Richard Samuel, *Science and Religion in Seventeenth-Century England* (Archon Books, Hamden, 1958)
White, Andrew Dickson, *A History of the Warfare of Science with Theology in Christendom*, 2 vols (1896; Dover, New York, 1960)

History
Ceram, C.W. [*pseud.* Marek, Kurt Willhelm], (trans Garside, E.B.), *Gods, Graves and Scholars* (Victor Gollancz, London, 1955)
Gay, Peter, *The Enlightenment* (Norton, New York, 1996)
Gibbon, Edward, *Decline and Fall of the Roman Empire*, 6 vols (London, 1776–87)
Hall, A. Rupert, *From Galileo to Newton 1630–1720* (Collins, London, 1963)
Hampson, Norman, *The Enlightenment* (Pelican, 1968)
Plumb, J.H., *England in the Eighteenth Century* (Penguin, 1950)

Index